Birkhäuser

Tutorials, Schools, and Workshops in the Mathematical Sciences

This series will serve as a resource for the publication of results and developments presented at summer or winter schools, workshops, tutorials, and seminars. Written in an accessible style, they present important and emerging topics in scientific research for PhD students and researchers. Filling a gap between traditional lecture notes, proceedings, and standard textbooks, the titles included in this series present material from the forefront of research.

Manuscripts are solicited by the editorial boards of each volume and then reviewed by a minimum of three peer reviewers to ensure the highest standards of scientific literature.

Baltic Institute of Mathematics Collection

Since 2012, the Baltic Institute of Mathematics has been organizing scientific events such as conferences, workshops, seminars, and schools. The main goal of these events is to facilitate networking between young and senior researchers to discuss the state-of-the-art and outstanding new results in mathematics, physics, and computer science, and their use in industry. The volumes in this collection are based on the Institute's events and are written in an accessible, pedagogical style for graduate students and early-career researchers.

Maria Ulan · Noémie C. Combe
Editors

Geometric Methods in Physical Systems: From Differentiable Structures to Applications

The Wisła 22 Winter School and Workshop

Birkhäuser

Editors
Maria Ulan
Baltic Institute of Mathematics
Warsaw, Poland

Noémie C. Combe
Baltic Institute of Mathematics
Warsaw, Poland

ISSN 2522-0969 ISSN 2522-0977 (electronic)
Tutorials, Schools, and Workshops in the Mathematical Sciences
ISBN 978-3-032-00398-0 ISBN 978-3-032-00399-7 (eBook)
https://doi.org/10.1007/978-3-032-00399-7

Mathematics Subject Classification: 76-06, 53-06, 70-06

This book is published under the imprint Birkhäuser, www.birkhauser-science.com by the registered company Springer Nature Switzerland AG
The registered company address is: Gewerbestrasse 11, 6330 Cham, Switzerland

Preface

This volume presents selected materials from the Winter School and Workshop Wisla 22 organized by the Baltic Institute of Mathematics. It brings together four complementary works that illustrate the power of geometric methods in understanding complex physical systems. The chapters progress from foundational mathematical structures to concrete applications in fluid dynamics and mechanical systems, highlighting the profound connections between differential geometry and physical phenomena.

We begin with fundamental mathematical concepts in Chapter "Classification of Differentiable Structures on the Non-hausdorff Line with Two Origins", which classifies differentiable structures on a non-Hausdorff line with two origins. This investigation of mathematical structures sets the stage for the applications that follow, demonstrating how geometric classifications can reveal surprising properties like the existence of uncountably many non-diffeomorphic structures.

Chapter "Generalized Geometry of 2D Incompressible Fluid Flows" transitions to fluid mechanics through a study of generalized geometry in two-dimensional incompressible fluid flows. By developing a family of generalized almost structures associated with Monge-Ampère equations, this work establishes the mathematical framework needed for analyzing fluid systems through geometric lenses.

Building on these foundations, Chapter "Nonlinear Differential Equations of Fluid Mechanics: Symmetries, Integrability, Singularities" expands this perspective with a comprehensive treatment of nonlinear differential equations in fluid mechanics. Here, the concepts of contact and symplectic geometry illuminate singular properties of fluid dynamics solutions, including multivalued solutions, caustics, and shock wave fronts across both barotropic and fully compressible Euler's equations.

Finally, Chapter "On Geometry of Mechanical Systems with Smooth and Non-smooth Nonholonomic Constraints" demonstrates how geometric methods extend beyond fluid mechanics to mechanical systems with nonholonomic constraints. This exploration of both smooth and non-smooth constraints reveals how geometric formulations can address challenging phenomena like discontinuities, collisions, and the counterintuitive stabilization of inverted pendulums.

Together, these chapters showcase how geometric approaches provide not only elegant mathematical frameworks but also practical insights into complex physical phenomena. From pure classification problems to applied analysis of pendulum dynamics, this collection demonstrates the unifying power of geometric thinking across diverse domains of mathematical physics.

By presenting these works in sequence, we hope to guide readers from abstract mathematical concepts to concrete physical applications, illustrating the remarkable ability of geometric methods to simplify, clarify, and unify our understanding of physical systems.

Warsaw, Poland
February 2025

Maria Ulan
Noémie C. Combe

Acknowledgements

The Baltic Institute of Mathematics owes its deepest gratitude to Jerzy Smit, whose boundless energy and commitment were crucial in bringing all the pieces together. Without his vision and dedication, the Wisla 22 Winter School and Workshop, and this book, would have remained merely an idea.

We also extend our sincere appreciation to Prof. Olav Arnfinn Laudal for kindliness, expertise and continuous support. Professor Jerzy Buzek, MEP deserves a special thank you for the honorary patronage of the Futurum 2020s Initiative that helps to organize and provide excellent meeting facilities during the Wisła 22 Winter School and Workshop.

The collaborative environment and stimulating discussions during this event were instrumental in refining the ideas explored throughout these chapters. The organizers would like to thank speakers of the lectures, and all participants for a great, friendly atmosphere and strong motivation to learn. Extra acknowledgments go to the authors for contributing and presenting their research.

We are indebted to the reviewers for their valuable comments and suggestions on preliminary drafts of the book. Their feedback has enhanced both the clarity and rigor of these works. The editors would like to thank the Birkhäuser Mathematics, Springer Nature crew, especially Chris Tominich for assistance throughout the publication process.

We acknowledge the participation of graduate students and early-career researchers whose questions and insights during the workshop sessions helped sharpen the focus of these presentations. Their enthusiasm for geometric approaches to physical problems gives us confidence in the future of this field.

Acknowledgements

Contents

Classification of Differentiable Structures on the Non-hausdorff Line with Two Origins 1
Mykola Lysynskyi and Sergiy Maksymenko

Generalized Geometry of 2D Incompressible Fluid Flows 39
Radek Suchánek

Nonlinear Differential Equations of Fluid Mechanics: Symmetries, Integrability, Singularities 61
Michael Roop

On Geometry of Mechanical Systems with Smooth and Non-smooth Nonholonomic Constraints 109
Maria Ulan

Contributors

Mykola Lysynskyi Department of Algebra and Topology, Institute of Mathematics of NAS of Ukraine, Kyiv, Ukraine

Sergiy Maksymenko Department of Algebra and Topology, Institute of Mathematics of NAS of Ukraine, Kyiv, Ukraine

Michael Roop Department of Mathematical Sciences, Chalmers University of Technology and University of Gothenburg, Gothenburg, Sweden

Radek Suchánek Department of Mathematics and Statistics, Masaryk University, Brno, Czech Republic

Maria Ulan Baltic Institute of Mathematics, Warsaw, Poland

Classification of Differentiable Structures on the Non-hausdorff Line with Two Origins

Mykola Lysynskyi and Sergiy Maksymenko

Abstract We classify differentiable structures on a line $\mathbb{L}$ with two origins being a non-Hausdorff but T_1 one-dimensional manifold obtained by "splitting" $0 \in \mathbb{R}$ into two points 0 and $\bar{0}$. For $k \subset \mathbb{N} \cup \{\infty\}$ let H be the group of homeomorphisms h of $\mathbb{R}$ such that $h(0) = 0$ and the restriction of h to $\mathbb{R} \setminus 0$ is a C^k-diffeomorphism. Let also D be the subgroup of H consisting of C^k-diffeomorphisms of $\mathbb{R}$ also fixing 0. It is shown that there is a natural bijection between C^k-structures on $\mathbb{L}$ (up to a C^k-diffeomorphism fixing both origins 0 and $\bar{0}$) and double D-coset classes $D \setminus H / D = \{DhD \mid h \in H\}$. Moreover, the set of all C^k-structures on $\mathbb{L}$ (up to a C^k-diffeomorphism which may also exchange origins) are in one-to-one correspondence with the set of double $(D, \pm)$-coset classes $D \setminus H^{\pm} / D = \{DhD \cup Dh^{-1}D \mid h \in H\}$. In particular, in contrast to the real line $\mathbb{R}$, the line with two origins $\mathbb{L}$ admits uncountably many pair-wise non-diffeomorphic C^k-structures for each $k = 1, 2, \ldots, \infty$.

1 Introduction

The present paper is devoted to some steps toward the study and classification of smooth structures on non-Hausdorff manifolds of dimension 1.

It is well known that each Hausdorff manifold M of dimension $\dim M \leq 3$ has a unique smooth structure up to a diffeomorphism. For $\dim M = 1$ this is classical result, and for $\dim M = 2, 3$ see Munkres [25, Theorem 6.3] and Whitehead [39, Corollary 1.18]. Stallings [34] shown that $\mathbb{R}^n$ has a unique C^∞ structure for $n \geq 5$, see also [3, Sect. 10]. Also, in the book Kirby and Siebenmann [19, Theorem on p. 155] it is proved that on every closed topological Hausdorff manifold of dimension ≥ 5 there exists only finitely many classes of non-diffeomorphic smooth structures.

M. Lysynskyi · S. Maksymenko (✉)
Department of Algebra and Topology, Institute of Mathematics of NAS of Ukraine, Kyiv, Ukraine
e-mail: maks@imath.kiev.ua

M. Lysynskyi
e-mail: m.lysynskyi@imath.kiev.ua

M. Ulan and N. C. Combe (eds.), *Geometric Methods in Physical Systems: From Differentiable Structures to Applications*, Tutorials, Schools, and Workshops in the Mathematical Sciences, https://doi.org/10.1007/978-3-032-00399-7_1

On the other hand, $\mathbb{R}^4$ has uncountably many pair-wise non-diffeomorphic C^∞ structures, see Freedman [6], Donaldson [2, Theorem 1], Freed and Uhlenbeck [5], Gompf [8] and Taubes [38]. It is also worth noting a result by Milnor [24] on the existence of distinct smooth structures on 7-sphere, as well as the papers Kervaire and Milnor [18] on the group of homotopy spheres, and Kervaire [17] on 10-manifold which does not admit any differentiable structures.

In contrast, differentiable structures of non-Hausdorff manifolds seem not to be essentially and systematically studied in the literature, see, e.g., the book by Gauld [7] and references therein. For instance, Haefliger and Reeb [9] proven existence of smooth structures on *non-Hausdorff letter* Y. Nyikos [26] studied smooth structures on the *long line* X, and showed that it admits non-diffeomorphic C^ω (real analytic) structures; however, it is not clear whether there are non-diffeomorphic C^∞-structures on X. The long line and the *Prüfer manifold* are also considered in [33, Appendix A]. Applications of non-Hausdorff manifolds to physics are discussed in [10, 14]. Also, in recent papers [27, 28] by D. O'Connel the structure of vector bundles over non-Hausdorff manifolds is also investigated.

In the present paper we will completely classify C^k-structures, $1 \leq k \leq \infty$, on another well-known non-Hausdorff one-dimensional manifold, called the *line with two origins* $\mathbb{L}$, and which can be described in (at least) the following three equivalent ways, see Fig. 1:

(1) Let $R = (\mathbb{R} \times 0) \sqcup (\mathbb{R} \times 1)$ be a disjoint union of two copies of $\mathbb{R}$. Then $\mathbb{L}$ is a quotient space of R obtained by identifying points $(t, 0)$ with $(t, 1)$ for $t \neq 0$.

(2) Let τ be the standard topology on $\mathbb{R}$. Then $\mathbb{L} = \mathbb{R} \sqcup \bar{0}$ is a disjoint union of $\mathbb{R}$ with some point $\bar{0}$ endowed with the following topology:

$$\eta = \tau \cup \{(W \setminus 0) \cup \{\bar{0}\} \mid 0 \in W \in \tau\},$$

whose elements are elements of τ and also open neighborhoods of 0 in which 0 is replaced with $\bar{0}$.

(3) Let $\mathcal{F}$ be the partition of $\mathbb{R}^2 \setminus 0$ into connected components of the intersections of $\mathbb{R}^2 \setminus 0$ with vertical lines $\{x = c\}_{c \in \mathbb{R}}$. Then $\mathbb{L}$ is the space of leaves $(\mathbb{R}^2 \setminus 0)/\mathcal{F}$ of that foliation (with the corresponding quotient topology), so that the

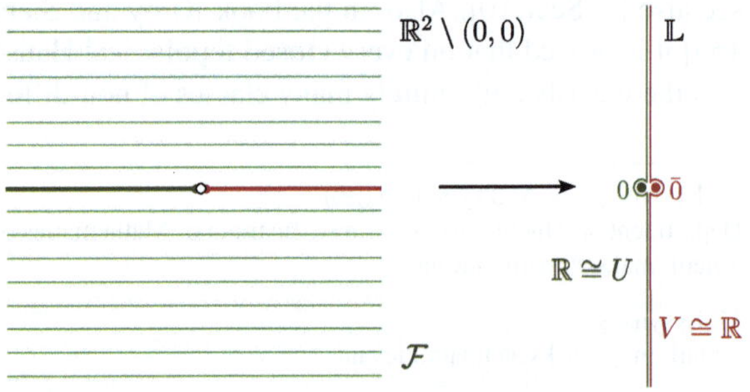

Fig. 1 Line with two origins $\mathbb{L}$

leaves corresponding to non-distinguishable points 0 and $\bar{0}$ are the open intervals $(-\infty; 0) \times 0$ and $(0; \infty) \times 0$.

It will be convenient to exploit the second model for $\mathbb{L}$. As in Fig. 1 denote

$$U := \mathbb{L} \setminus \bar{0} = \mathbb{R}, \qquad V := \mathbb{L} \setminus 0 = (\mathbb{R} \setminus 0) \cup \bar{0}. \tag{1}$$

Lemma 1 *$\mathbb{L}$ is locally Euclidean topological space being also, one-dimensional, second countable, T_1 but not T_2.* □

Proof This lemma is trivial, and we will only explain the non-T_2 property. Note that U is an open neighborhood of 0 which does not contain $\bar{0}$, while V is an open neighborhood of 0 which does not contain 0. On the other hand, by the definition of topology η, any pair of neighborhoods of 0 and $\bar{0}$ intersects. Thus, $\mathbb{L}$ is T_1 but not T_2. □

Our main result is an explicit description of

- the set of differentiable C^k-structures on $\mathbb{L}$ up to a C^k-diffeomorphism and
- the sets of C^k-diffeomorphisms between C^k-structures on $\mathbb{L}$, see Theorem 1.

As a consequence we will show that, in contrast with the real line $\mathbb{R}$, the line $\mathbb{L}$ with two origins admits uncountably many pair-wise non-diffeomorphic structures, see Theorem 2. The exposition is intended to be elementary and self-contained and is aimed to be available for a large audience of readers interested in these topics.

1.1 Notation

For each $k \in \mathbb{N} \cup \{\infty\}$ define the following groups:

- $\mathcal{H}(\mathbb{R}, 0)$ is the group of homeomorphisms of $\mathbb{R}$ fixing 0.
- $\mathcal{H}^+(\mathbb{R}, 0)$ is the normal subgroup of $\mathcal{H}(\mathbb{R}, 0)$ (of index 2) consisting of orientation preserving homeomorphisms, and $\mathcal{H}^-(\mathbb{R}, 0) = \mathcal{H}(\mathbb{R}, 0) \setminus \mathcal{H}^+(\mathbb{R}, 0)$ is the another adjacent class of $\mathcal{H}^+(\mathbb{R}, 0)/\mathcal{H}(\mathbb{R}, 0)$ consisting of homeomorphisms reversing orientation.
- $\mathcal{W}^k(\mathbb{R}, 0)$ is the subgroup of $\mathcal{H}(\mathbb{R}, 0)$ consisting of homeomorphisms h such that $h(0) = 0$ and the restriction $h|_{\mathbb{R}\setminus 0} : \mathbb{R} \setminus 0 \to \mathbb{R} \setminus 0$ is a C^k-diffeomorphism of $\mathbb{R} \setminus 0$.
- $\mathrm{Diff}^k(\mathbb{R}, 0)$ is the subgroup of $\mathcal{W}^k(\mathbb{R}, 0)$ consisting of C^k-diffeomorphisms of $\mathbb{R}$.
- $\mathcal{E}^k(\mathbb{R}, 0)$ is the subgroup of $\mathrm{Diff}^k(\mathbb{R}, 0)$ consisting of diffeomorphisms h such that $h^{(i)}(0) = 0$ for $i = 2, 3, \dots, k$. Notice that if k is finite, then every $h \in \mathcal{E}^k(\mathbb{R}, 0)$ is given by the formula $h(x) = \tau x + x^k g(x)$ for some $\tau \neq 0$ and a continuous function $g : \mathbb{R} \to \mathbb{R}$ being C^k on $\mathbb{R} \setminus 0$. On the other hand, if $k = \infty$, then $h(x) = \tau x + g(x)$ for some $\tau \neq 0$ and a C^∞ function being *flat* at 0, i.e., $g^{(i)}(0) = 0$ for all $i \in \mathbb{N}$. In particular, $\mathcal{E}^1(\mathbb{R}, 0) = \mathrm{Diff}^1(\mathbb{R}, 0)$.

- $\mathrm{Diff}^k_{\mathrm{nb}}(\mathbb{R}, 0)$ is the group of C^k-diffeomorphisms fixed on some neighborhood of 0.
- Also for $* \in \{\pm\}$ and $\mathcal{X} \in \{\mathcal{W}, \mathrm{Diff}, \mathcal{E}\}$ put $\mathcal{X}^{k,*}(\mathbb{R}, 0) := \mathcal{H}^*(\mathbb{R}, 0) \cap \mathcal{X}^k(\mathbb{R}, 0)$.

Evidently, $\mathrm{Diff}^k_{\mathrm{nb}}(\mathbb{R}, 0) \subset \mathcal{E}^{k,+}(\mathbb{R}, 0)$ and for all $k' < k$ we have the following inclusions

$$\begin{array}{ccccc} \mathcal{E}^k(\mathbb{R},0) & \hookrightarrow & \mathrm{Diff}^k(\mathbb{R},0) & \hookrightarrow & \mathcal{W}^k(\mathbb{R},0) \\ \downarrow & & \downarrow & & \downarrow \\ \mathcal{E}^{k'}(\mathbb{R},0) & \hookrightarrow & \mathrm{Diff}^{k'}(\mathbb{R},0) & \hookrightarrow & \mathcal{W}^k(\mathbb{R},0)[k']. \end{array}$$

Example 1 For any real numbers $\alpha, \beta \neq 0$ having the same sign, and $s, t > 0$ the following homeomorphism

$$h\colon \mathbb{R} \to \mathbb{R}, \qquad h(x) = \begin{cases} \alpha(-x)^s, & x < 0, \\ \beta x^t, & x \geq 0 \end{cases}$$

belongs to $\mathcal{W}^k(\mathbb{R}, 0)$. Evidently, $h \in \mathrm{Diff}^k(\mathbb{R}, 0)$ iff $\alpha = \beta$ and $t = s = 1$, i.e., when h is linear, and in that case it belongs even to $\mathcal{E}^\infty(\mathbb{R}, 0)$. □

Consider the following *left* action of the product $\mathrm{Diff}^k(\mathbb{R}, 0) \times \mathrm{Diff}^k(\mathbb{R}, 0)$ on $\mathcal{W}^k(\mathbb{R}, 0)$ defined by

$$\mu\colon \mathrm{Diff}^k(\mathbb{R}, 0) \times \mathrm{Diff}^k(\mathbb{R}, 0) \times \mathcal{W}^k(\mathbb{R}, 0) \to \mathcal{W}^k(\mathbb{R}, 0), \qquad \mu(a, b, h) = a \circ h \circ b^{-1}.$$

The corresponding orbit space of $\mathcal{W}^k(\mathbb{R}, 0)$ is called the *set of* $\mathrm{Diff}^k(\mathbb{R}, 0)$*-double cosets* and denoted by

$$\mathrm{Diff}^k(\mathbb{R}, 0) \setminus \mathcal{W}^k(\mathbb{R}, 0) \,/\, \mathrm{Diff}^k(\mathbb{R}, 0).$$

One easily checks, see Lemma 4, that $g, h \in \mathcal{W}^k(\mathbb{R}, 0)$ belong to the same orbit of the above action iff $\mathrm{Diff}^k(\mathbb{R}, 0) \cap h\,\mathrm{Diff}^k(\mathbb{R}, 0) g^{-1} \neq \varnothing$.

Also, consider a partition of $\mathcal{W}^k(\mathbb{R}, 0)$ into larger sets. Say that $g, h \in \mathcal{W}^k(\mathbb{R}, 0)$ are $(\mathrm{Diff}^k(\mathbb{R}, 0), \pm)$-equivalent, if *either g or g^{-1} belong to the orbit of h with respect to the above action μ*. One easily checks that this relation is an equivalence, see Sect. 2.4. The corresponding equivalence classes will be called $(\mathrm{Diff}^k(\mathbb{R}, 0), \pm)$-*double cosets*, and the set of all such classes will be denoted by

$$\mathrm{Diff}^k(\mathbb{R}, 0) \setminus \mathcal{W}^k(\mathbb{R}, 0)^{\pm 1} \,/\, \mathrm{Diff}^k(\mathbb{R}, 0).$$

It can also be defined as the orbit set of a certain natural action of the *wreath product* $\mathrm{Diff}^k(\mathbb{R}, 0) \wr \mathbb{Z}_2$ extending the action μ, see Sect. 2.4.

1.2 Main Results

Let $\mathcal{H}(\mathbb{L})$ be the group of all homeomorphisms of $\mathbb{L}$. It is easy to show that every homeomorphism h of $\mathbb{L}$ either fixes or exchanges points 0 and $\bar{0}$, see Lemma 5. In particular, the restriction of h to the complement $\mathbb{L} \setminus \{0, \bar{0}\} = \mathbb{R} \setminus 0$ is also invariant under h. Moreover, $h|_{\mathbb{R}\setminus 0} \colon \mathbb{R} \setminus 0 \to \mathbb{R} \setminus 0$ is either strictly increasing or strictly decreasing homeomorphism. Therefore we will also say that h *preserves* orientation of $\mathbb{L}$ if $h|_{\mathbb{R}\setminus 0}$ is increasing, and h *reverses* orientation of $\mathbb{L}$ otherwise.

For $* \in \{+, -, \text{fix}, \text{ex}\}$ denote by $\mathcal{H}^*(\mathbb{L})$ the *subset* of $\mathcal{H}(\mathbb{L})$ consisting of homeomorphisms respectively *preserving* or *reversing* orientation of $\mathbb{L}$, or *fixing* or *exchanging* 0 and $\bar{0}$. Also for $* \in \{+, -\}$ and $\bullet \in \{\text{fix}, \text{ex}\}$ we put

$$\mathcal{H}^{*,\bullet}(\mathbb{L}) := \mathcal{H}^*(\mathbb{L}) \cap \mathcal{H}^\bullet(\mathbb{L}).$$

Evidently, $\mathcal{H}^{\text{fix}}(\mathbb{L})$ and $\mathcal{H}^+(\mathbb{L})$ are normal subgroups of $\mathcal{H}(\mathbb{L})$ of index 2, while their intersection $\mathcal{H}^{+,\text{fix}}(\mathbb{L})$ has index 4.

Fix once and for all some $k \in \mathbb{N} \cup \{\infty\}$. Given two C^k-structures $\mathfrak{A}$ and $\mathfrak{B}$ on $\mathbb{L}$ let:

- $\text{Diff}_{\mathfrak{A},\mathfrak{B}}(\mathbb{L})$ be the set of all C^k-diffeomorphisms of $\mathbb{L}$ from $\mathfrak{A}$ to $\mathfrak{B}$;
- $\text{Diff}^*_{\mathfrak{A},\mathfrak{B}}(\mathbb{L}) = \text{Diff}_{\mathfrak{A},\mathfrak{B}}(\mathbb{L}) \cap \mathcal{H}^*(\mathbb{L})$, where $* \in \{-, +, \text{fix}, \text{ex}\}$;
- $\text{Diff}^{*,\bullet}_{\mathfrak{A},\mathfrak{B}}(\mathbb{L}) = \text{Diff}^*_{\mathfrak{A},\mathfrak{B}}(\mathbb{L}) \cap \text{Diff}^\bullet_{\mathfrak{A},\mathfrak{B}}(\mathbb{L}) = \text{Diff}_{\mathfrak{A},\mathfrak{B}}(\mathbb{L}) \cap \mathcal{H}^{*,\bullet}(\mathbb{L})$, where $* \in \{-, +\}$, and $\bullet \in \{\text{fix}, \text{ex}\}$.

Notice that though we omitted k from notations, it is implicitly determined by the notations of C^k-structures $\mathfrak{A}$ and $\mathfrak{B}$.

If $\mathfrak{A} = \mathfrak{B}$, then we will abbreviate $\text{Diff}^*_{\mathfrak{A},\mathfrak{B}}(\mathbb{L})$ and $\text{Diff}^{*,\bullet}_{\mathfrak{A},\mathfrak{B}}(\mathbb{L})$ to $\text{Diff}^*_{\mathfrak{A}}(\mathbb{L})$ and $\text{Diff}^{*,\bullet}_{\mathfrak{A}}(\mathbb{L})$ respectively. For instance, $\text{Diff}^+_{\mathfrak{A}}(\mathbb{L})$, resp. $\text{Diff}^{\text{fix}}_{\mathfrak{A}}(\mathbb{L})$, is the *group* of C^k-diffeomorphisms of $\mathfrak{A}$ preserving orientation, resp. fixing 0 and $\bar{0}$, while $\text{Diff}^{+,\text{fix}}_{\mathfrak{A}}(\mathbb{L})$ is their intersection.

Evidently, we have the following partition

$$\text{Diff}_{\mathfrak{A},\mathfrak{B}}(\mathbb{L}) = \text{Diff}^{+,\text{fix}}_{\mathfrak{A},\mathfrak{B}}(\mathbb{L}) \sqcup \text{Diff}^{-,\text{fix}}_{\mathfrak{A},\mathfrak{B}}(\mathbb{L}) \sqcup \text{Diff}^{+,\text{ex}}_{\mathfrak{A},\mathfrak{B}}(\mathbb{L}) \sqcup \text{Diff}^{-,\text{ex}}_{\mathfrak{A},\mathfrak{B}}(\mathbb{L}),$$

however depending on $\mathfrak{A}$ and $\mathfrak{B}$ some of $\text{Diff}^{*,\bullet}_{\mathfrak{A},\mathfrak{B}}(\mathbb{L})$ may be empty.

The following theorem classifies C^k-structures on $\mathbb{L}$. In fact, we will prove more detailed and explicit statements describing also the sets of diffeomorphisms between C^k-structures fixing or exchanging 0 and $\bar{0}$, see Theorem 5.

Theorem 1 *Let* $k \in \mathbb{N} \cup \{\infty\}$. *Then*

- C^k*-structures on* $\mathbb{L}$ up to a C^k-diffeomorphism *are in one-to-one correspondence with the set* $\text{Diff}^k(\mathbb{R}, 0) \setminus \mathcal{W}^k(\mathbb{R}, 0)^{\pm 1} \,/\, \text{Diff}^k(\mathbb{R}, 0)$ *of* $(\text{Diff}^k(\mathbb{R}, 0), \pm)$*-double coset classes;*

- *while* C^k*-structures on* $\mathbb{L}$ up to a C^k-diffeomorphism fixing 0 and $\bar{0}$ *are in one-to-one correspondence with the set* $\mathrm{Diff}^k(\mathbb{R},0) \setminus \mathcal{W}^k(\mathbb{R},0) \,/\, \mathrm{Diff}^k(\mathbb{R},0)$ *of* $\mathrm{Diff}^k(\mathbb{R},0)$*-double coset classes.* □

The following lemma coincides with statements (3) and (4) of Theorem 4.

Lemma 2 *For any* C^k*-structure,* $\mathrm{Diff}^{+,\mathrm{fix}}_{\mathfrak{A}}(\mathbb{L})$ *contains a normal subgroup isomorphic to the group* $\mathrm{Diff}^k_{\mathrm{nb}}(\mathbb{R},0)$ *of diffeomorphisms of* $\mathbb{R}$ *fixed near* 0*. In particular,* $\mathrm{Diff}^{+,\mathrm{fix}}_{\mathfrak{A}}(\mathbb{L})$ *is always uncountable.*

Even more, if $\mathfrak{A}$ *and* $\mathfrak{B}$ *are* C^k*-structures on* $\mathbb{L}$ *and* $* \in \{\mathrm{fix}, \mathrm{ex}\}$ *are such that* $\mathrm{Diff}^{*,\bullet}_{\mathfrak{A},\mathfrak{B}}(\mathbb{L})$ *is non-empty, then* $\mathrm{Diff}^{*,\bullet}_{\mathfrak{A},\mathfrak{B}}(\mathbb{L})$ *is in fact uncountable.* □

Our last statement (see Theorem 2) presents examples of C^k-structures for which either of the sets $\mathrm{Diff}^{\mathrm{fix}}_{\mathfrak{A},\mathfrak{B}}(\mathbb{L})$ and $\mathrm{Diff}^{\mathrm{ex}}_{\mathfrak{A},\mathfrak{B}}(\mathbb{L})$ is empty or non-empty.

A C^k-atlas A on $\mathbb{L}$ will be called *minimal* if it consists of only two charts $\varphi\colon U \to \mathbb{R}$ and $\psi\colon V \to \mathbb{R}$ being homeomorphisms onto and such that $\varphi(0) = \psi(\bar{0}) = 0$. We will show in Lemma 16 that every C^k-structure on $\mathbb{L}$ admits such a minimal atlas.

In particular, let $\mathrm{id}_u\colon U \to \mathbb{R}$ be the identity map, and $\mathrm{id}_v\colon V \to \mathbb{R}$ be the homeomorphism given by $\mathrm{id}_v = \mathrm{id}_u$ on $V \setminus \{0\}$ and $\mathrm{id}_v(\bar{0}) = 0$. Then for each $g \in \mathcal{W}^k(\mathbb{R},0)$ one can define the following minimal atlas $\mathsf{S}(g) = \{(U, \mathrm{id}_u), (V, g \circ \mathrm{id}_v)\}$. Such an atlas will be called *special minimal for* g, see Sect. 5.5. Evidently, its transition map is

$$g \circ \mathrm{id}_v \circ \mathrm{id}_u^{-1}|_{\mathbb{R}\setminus 0} = g|_{\mathbb{R}\setminus 0},$$

whence $\mathsf{S}(g)$ is C^k.

Further, for each $a > 0$ define the following homeomorphism $w_a \in \mathcal{W}^k(\mathbb{R},0)$ by

$$w_a(x) = \begin{cases} x, & x \le 0, \\ ax, & x > 0, \end{cases} \tag{2}$$

see Fig. 2.

It is a particular case of the homeomorphism h from Example 1 for $\alpha = 1$, $\beta = a$, $s = t = 1$. Let $\mathfrak{S}(a)$ be the corresponding C^k-structure defined by the minimal atlas $\mathsf{S}(w_a)$.

Fig. 2 Homeomorphism w_a

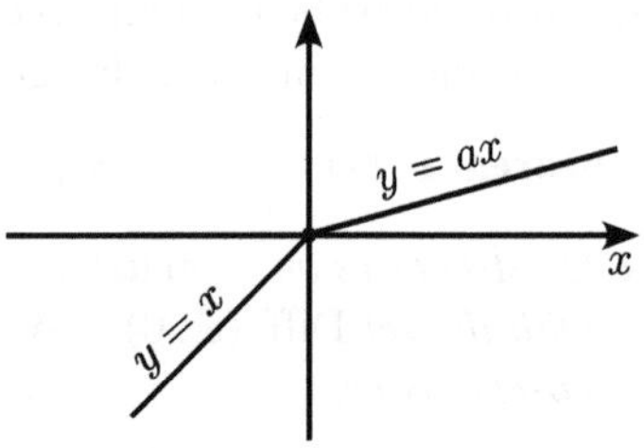

Theorem 2 *(1) Suppose $a = 1$, so $w_a = \mathrm{id}_{\mathbb{R}}$. Then there is an isomorphism*

$$\eta\colon \mathrm{Diff}_{\mathfrak{S}(1)}(\mathbb{L}) \cong \mathrm{Diff}^k(\mathbb{R}, 0) \times \mathbb{Z}_2,$$

such that for $* \in \{\pm\}$

$$\eta\big(\mathrm{Diff}^{*,\mathrm{fix}}_{\mathfrak{S}(1)}(\mathbb{L})\big) = \mathrm{Diff}^{k,*}(\mathbb{R}, 0) \times \{1\}, \quad \eta\big(\mathrm{Diff}^{*,\mathrm{ex}}_{\mathfrak{S}(1)}(\mathbb{L})\big) = \mathrm{Diff}^{k,*}(\mathbb{R}, 0) \times \{-1\}. \tag{3}$$

(2) Let $a \neq 1$. Then $\mathrm{Diff}^{+,\mathrm{ex}}_{\mathfrak{S}(a)}(\mathbb{L}) = \mathrm{Diff}^{-,\mathrm{fix}}_{\mathfrak{S}(a)}(\mathbb{L}) = \varnothing$, and $\mathrm{Diff}_{\mathfrak{S}(a)}(\mathbb{L}) = \mathrm{Diff}^{+,\mathrm{fix}}_{\mathfrak{S}(a)}(\mathbb{L}) \sqcup \mathrm{Diff}^{-,\mathrm{ex}}_{\mathfrak{S}(a)}(\mathbb{L})$, so every self-diffeomorphism of $\mathfrak{S}(a)$ either fixes 0 and $\bar{0}$ and preserves *orientation, or exchanges 0 and $\bar{0}$ and* reverses *orientation. Moreover, there exists an isomorphism*

$$\eta\colon \mathrm{Diff}^{+,\mathrm{fix}}_{\mathfrak{S}(a)}(\mathbb{L}) \cong \mathcal{E}^{k,+}(\mathbb{R}, 0),$$

and an element $p \in \mathrm{Diff}^{-,\mathrm{ex}}_{\mathfrak{S}(a)}(\mathbb{L})$ of order 2. Hence, $\mathrm{Diff}_{\mathfrak{S}(a)}(\mathbb{L})$ splits into a semidirect product $\mathrm{Diff}^{+,\mathrm{fix}}_{\mathfrak{S}(a)}(\mathbb{L}) \rtimes \langle p\rangle$, and thus it is isomorphic to a certain[1] *semidirect product $\mathcal{E}^{k,+}(\mathbb{R}, 0) \rtimes \mathbb{Z}_2$.*

(3) Also, for each $a \neq 1$ the structures $\mathfrak{S}(a)$ and $\mathfrak{S}(1/a)$ are C^k-diffeomorphic. In fact, $\mathrm{Diff}^{+,\mathrm{fix}}_{\mathfrak{S}(a),\mathfrak{S}(1/a)}(\mathbb{L}) = \mathrm{Diff}^{-,\mathrm{ex}}_{\mathfrak{S}(a),\mathfrak{S}(1/a)}(\mathbb{L}) = \varnothing$ and

$$\mathrm{Diff}_{\mathfrak{S}(a),\mathfrak{S}(1/a)}(\mathbb{L}) = \mathrm{Diff}^{+,\mathrm{ex}}_{\mathfrak{S}(a),\mathfrak{S}(1/a)}(\mathbb{L}) \sqcup \mathrm{Diff}^{-,\mathrm{fix}}_{\mathfrak{S}(a),\mathfrak{S}(1/a)}(\mathbb{L}),$$

so every C^k-diffeomorphism from $\mathfrak{S}(a)$ to $\mathfrak{S}(1/a)$ either fixes 0 and $\bar{0}$ and reverses *orientation, or exchanges 0 and $\bar{0}$ and* preserves *orientation. Moreover, for every $h \in \mathrm{Diff}^{-,\mathrm{fix}}_{\mathfrak{S}(a),\mathfrak{S}(1/a)}(\mathbb{L})$ and $g \in \mathrm{Diff}^{+,\mathrm{ex}}_{\mathfrak{S}(a),\mathfrak{S}(1/a)}(\mathbb{L})$ we have the following bijections:*

$$\mathrm{Diff}^{+,\mathrm{ex}}_{\mathfrak{S}(a),\mathfrak{S}(1/a)}(\mathbb{L}) \xrightarrow{k \mapsto g^{-1}\circ k} \mathrm{Diff}^{+,\mathrm{fix}}_{\mathfrak{S}(a)}(\mathbb{L}) \xrightarrow{k \mapsto h\circ k} \mathrm{Diff}^{-,\mathrm{fix}}_{\mathfrak{S}(a),\mathfrak{S}(1/a)}(\mathbb{L}).$$

(4) Finally, if $a \neq b, 1/b$, then $\mathrm{Diff}_{\mathfrak{S}(a),\mathfrak{S}(b)}(\mathbb{L}) = \varnothing$, i.e., these C^k-structures are not diffeomorphic. In particular, $\mathbb{L}$ has uncountably many pair-wise non-diffeomorphic C^k-structures of each $k \in \mathbb{N} \cup \infty$. □

Remark Notice that we have a surjective homomorphism $j^1\colon \mathcal{E}^{k,+}(\mathbb{R}, 0) \to \mathbb{R} \setminus 0$, $j^1(h) = h'(0)$. Let G_k be its kernel. Then G_k consists of diffeomorphisms k-tangent

[1] Notice that $\mathcal{E}^{k,-}(\mathbb{R}, 0)$ also contains the diffeomorphism $q\colon \mathbb{R} \to \mathbb{R}$, $q(x) = -x$, of order 2, and therefore $\mathcal{E}^k(\mathbb{R}, 0)$ also splits into the semidirect product $\mathcal{E}^{k,+}(\mathbb{R}, 0) \rtimes \langle q\rangle$. However, *it is not clear* whether there is an isomorphism between $\mathrm{Diff}_{\mathfrak{S}(a)}(\mathbb{L}) \cong \mathrm{Diff}^{+,\mathrm{fix}}_{\mathfrak{S}(a)}(\mathbb{L}) \rtimes \langle p\rangle$ and $\mathcal{E}^k(\mathbb{R}, 0) \cong \mathcal{E}^{k,+}(\mathbb{R}, 0) \rtimes \langle q\rangle$ extending η.

to the identity and was studied in [35]. Moreover, the group G_∞ plays an important role in the theory of foliations, see [16, 32, 36]. □

Structure of the Paper

Section 2 is preliminary: it contains definitions of double cosets and branch points of non-Hausdorff spaces. In Sect. 3 we discuss differentiable structures on locally Euclidean spaces not assuming them to be manifolds in a usual sense, i.e., Hausdorff and second countable. In Sect. 4 we present a proof of a well-known result that any two C^k-structures, $1 \leq k \leq \infty$, on the real line are diffeomorphic, see Corollary 1.1. In fact we need a more explicit (and actually equivalent) statement claiming that if M is a one-dimensional locally Euclidean space endowed with some C^k-structure and $U \subset M$ is an open subset homeomorphic with $\mathbb{R}$, then there exists a C^k-compatible chart $\varphi\colon U \to \mathbb{R}$ being a homeomorphism onto, see Corollary 1. In Sect. 5 we classify C^k-structures on $\mathbb{L}$, and, in particular, prove Theorem 1, see Theorem 5. Finally in Sect. 6 we prove Theorem 2.

2 Preliminaries

Throughout the paper the arrows $\hookrightarrow$ and $\twoheadrightarrow$ mean *monomorphism* and *epimorphism* respectively. We will also regard the cyclic group $\mathbb{Z}_2$ of order 2 as the multiplicative group $\{\pm 1\}$.

2.1 Semidirect Products

Let H be a group with a unit e and A be a subgroup of H. A homomorphism $s\colon H \to A$ is called *retraction* if $s(a) = a$ for all $a \in A$.

Suppose $s\colon H \to A$ is a homomorphism-retraction, and let K be its kernel. Then A acts on K by conjugations, and this allows to define the following group $K \rtimes_s A$ called the *semidirect product* of K and A corresponding to the retraction s. Namely, $K \rtimes_s A$ is a Cartesian product of *sets* $K \times A$ with the following multiplication:

$$(k, a)(l, b) := (kala^{-1}, ab), \qquad (k, a), (l, b) \in K \times A.$$

Moreover, the map $\gamma\colon K \rtimes_s A \to H$, $\gamma(k, a) = ka$, is an isomorphism of groups making commutative the following diagram:

$$\begin{array}{ccccc} K & \xrightarrow{k \mapsto (k,e)} & K \rtimes_s A & \xrightarrow{(k,a)\mapsto a} & A \\ \| & & \gamma \downarrow \cong & & \| \\ K & \longrightarrow & H & \xrightarrow{s} & A \end{array}$$

We will exploit this construction several times.

2.2 Wreath Product $D \wr \mathbb{Z}_2$

Let D be a group and

$$\phi \colon D \times D \to D \times D, \qquad \phi(a, b) = (b, a)$$

be the automorphism exchanging coordinates. It has the order 2, and therefore defines a monomorphism $\bar{\phi} \colon \mathbb{Z}_2 \to \mathrm{Aut}(D \times D)$, given by $\bar{\phi}(1) = \mathrm{id}_{D\times D}$ and $\bar{\phi}(-1) = \phi$.

The corresponding semidirect product $(D \times D) \rtimes \mathbb{Z}_2$ is called the *wreath product* of D and $\mathbb{Z}_2$ and denoted by $D \wr \mathbb{Z}_2$, see, e.g., [23]. Thus, $D \wr \mathbb{Z}_2$ is the direct product of *sets* $D \times D \times \mathbb{Z}_2$ with the following binary operation:

$$(a, b, \delta)(c, d, \varepsilon) = \begin{cases} (ac, bd, \delta\varepsilon), & \delta = +1, \\ (ad, bc, \delta\varepsilon), & \delta = -1. \end{cases}$$

2.3 Double Cosets $C \setminus H / D$

We recall here the definition of double cosets, see, e.g., [11, Sect. 1.7]. Let H be a group and C, D be two subgroups. Then we have a natural *left* action of the product $C \times D$ on H:

$$C \times D \times H \to H, \qquad (c, d) \cdot h = chd^{-1}.$$

The corresponding orbit of $h \in H$ is called the (C, D)-*double coset* of h and is denoted by ChD. If $C = D$, then DhD is also called the D-*double coset* of h.

Notice that

$$ChD = \{chd^{-1} \mid c \in C, d \in D\} = \{chd \mid c \in C, d \in D\} = \bigcup_{c\in C} chD = \bigcup_{d\in D} Chd.$$

In particular, it is the union of left cosets of D corresponding to elements of the right coset Ch, as well as the union of right cosets of C corresponding to elements of the left coset hD. The partition of H into (C, D)-double cosets is denoted by

$$C \setminus H / D.$$

Partitions into double cosets are not so "regular" as partitions into left or right cosets, and they are often hard to compute, see, e.g., [12, 13, 21, 31], and references therein. For instance, the double coset classes may have distinct sizes. The reason is that the above action of $C \times D$ on H is not always free.

Example 2 Let $\mathbb{D}_3 = \langle r, s \mid r^3 = s^2 = (rs)^2 = 1 \rangle$ be the dihedral group consisting of six elements e, r, r^2, s, sr, sr^2. Consider two subgroups $A = \{e, s\}$ and $B = \{e, sr\}$ of $\mathbb{D}_3$. One easily checks that

$$A \setminus \mathbb{D}_3 \,/\, B = \big\{\{e, s, r, sr\},\ \{r^2, sr^2\}\big\},$$
$$A \setminus \mathbb{D}_3 \,/\, A = A \setminus \mathbb{D}_3^{\pm 1} \,/\, A = \big\{\{e, s\}, \{r, r^2, sr, sr^2\}\big\}.$$

The following lemma is trivial. We will refer to it several times, and therefore it will be convenient to put it as a separate statement:

Lemma 3 *If $C = D$ and this group is normal in H, then $C \setminus H \,/\, D$ coincides with the partition H/D into (no matter left or right) cosets of D in H.*

Let us also mention one more description of $C \setminus H \,/\, D$ which will be useful in this paper.

Lemma 4 *Let $g, h \in H$. Then the following conditions are equivalent:*

(1) $g \in ChD$, which is the same as $CgD = ChD$;
(2) $C \cap gDh^{-1} \neq \varnothing$. □

Proof (1)⇒(2) Suppose $g \in ChD$, so $g = chd^{-1}$ for some $c \in C$ and $d \in D$. Then $c = gdh^{-1} \in C \cap gDh^{-1} \neq \varnothing$.

(2)⇒(1) Conversely, if $C \cap gDh^{-1}$ contains some element c, i.e., $c = gdh^{-1}$ for some $\partial \in D$, then $g = chd^{-1} \in ChD$. □

2.4 Double Cosets $D \setminus H^{\pm 1} \,/\, D$

Suppose $C = D$. Then for each $h \in H$, we have that $(DhD)^{-1} = Dh^{-1}D$ is the D-double coset of h^{-1}, see, e.g., the discussion in [4].

$$Dh^{\pm 1}D := DhD \cup Dh^{-1}D = \{chd, ch^{-1}d \mid c, d \in D\}$$

will be called *$(D, \pm)$-double coset of h*. Thus, if DhD mutually contains some $g \in H$ and its inverse, then $Dh^{\pm 1}D = DhD = Dh^{-1}D$. Otherwise, $Dh^{\pm 1}D = DhD \sqcup Dh^{-1}D$.

It follows that $(D, \pm)$-double cosets yield a partition on H which will be denoted by

$$D \setminus H^{\pm 1} \,/\, D.$$

One can give another description of $D \setminus H^{\pm 1} / D$. Note that we have a natural *left* action

$$(D \wr \mathbb{Z}_2) \times H \to H, \qquad (a, b, \delta) \cdot h := ah^{\delta} b^{-1}.$$

Indeed, if $a, b, c, d \in D$ and $h \in H$, then

$$\begin{aligned}
(a, b, 1)\big((c, d, 1) \cdot h\big) &= a(chd^{-1})b^{-1} = (ac, bd, 1) \cdot h = \big((a, b, 1)(c, d, 1)\big) \cdot h, \\
(a, b, -1)\big((c, d, 1) \cdot h\big) &= a(chd^{-1})^{-1}b^{-1} = (ad, bc, -1) \cdot h = \big((a, b, -1)(c, d, 1)\big) \cdot h, \\
(a, b, 1)\big((c, d, -1) \cdot h\big) &= ach^{-1}d^{-1}b^{-1} = (ac, bd, -1) \cdot h = \big((a, b, 1)(c, d, -1)\big) \cdot h, \\
(a, b, -1)\big((c, d, -1) \cdot h\big) &= a(ch^{-1}d^{-1})^{-1}b^{-1} = (ad, bc, 1) \cdot h = \big((a, b, -1)(c, d, -1)\big) \cdot h.
\end{aligned}$$

It is now easy to see that $D \setminus H^{\pm 1} / D$ coincides with the partition into orbits of that action of $D \wr \mathbb{Z}_2$.

2.5 *Branch Points*

Let M be a topological space with topology η, and $x \in M$ be a point. Denote by $\mathsf{hcl}(x) := \bigcap_{x \in W \in \eta} \overline{W}$ the intersection of closures of all open neighborhoods of x, we will call it the *Hausdorff closure of* x. Evidently, $x \in \mathsf{hcl}(x)$. Also note that $\mathsf{hcl}(x)$ consists of points $y \in M$ such that x and y have no disjoint neighborhoods.

Say that x is a *branch* point of M whenever $\{x\} \neq \mathsf{hcl}(x)$, i.e., $\mathsf{hcl}(x)$ consists of more that one point. It is well known and is easy to see that M is Hausdorff iff $\{x\} = \mathsf{hcl}(x)$ for all $x \in M$, i.e., M has no branch points. There are no standard terms for such points. For instance in [28] they are called *Hausdorff-violating*.

The following simple lemma is left for the reader.

Lemma 5 *Let* $h: M \to M$ *be a homeomorphism of* M. *Then* $h(\mathsf{hcl}(x)) = \mathsf{hcl}(h(x))$ *for all* $x \in M$. *In particular, the set of branch points of* M *is invariant under homeomorphisms.* □

3 Differentiable Structures on Locally Euclidean Spaces

This section briefly recalls definition of differentiable structures on locally Euclidean spaces. We refer the reader to the book by Lee [22, Chaps. 1 and 2], and also to the paper of Takens [37] for an equivalent approach.

Usually a *manifold* is a Hausdorff second countable locally Euclidean topological space. The first two conditions are necessary and sufficient for metrizability of manifolds and for the possibility to embed them into some Euclidean spaces. However, for defining differentiable structures only the property of being locally Euclidean is enough. Therefore we will assume here only that property. For simplicity we also do not consider manifolds with boundary.

For each $m \geq 0$ the real m-dimensional linear space $\mathbb{R}^m$ with the usual topology is called *Euclidean*. Note that $\mathbb{R}^0 = \{0\}$ is a one-point space being a 0-dimensional linear space. Also recall that a homeomorphism $h\colon U \to V$ between open subset of $\mathbb{R}^m$ is called a *C^k-diffeomorphism* for some $k \in \mathbb{N} \cup \{\infty\}$, if both h and h^{-1} are C^k-maps.

3.1 Locally Euclidean Spaces

Let M be a topological space. A *chart* on M is an open embedding (i.e., a homeomorphism onto open subset) $\varphi\colon U \to \mathbb{R}^m$ of some open subset $U \subset M$ for some $m \geq 0$. In this case we will say that this chart is *m-dimensional*, U is its *domain*, and φ is its *coordinate map*. It will often be convenient to regard such a chart as a pair $\mathsf{u} = (U, \varphi)$.

Notice that then for every open subset $U' \subset U$ the restriction $\varphi|_{U'}\colon U' \to \mathbb{R}^m$ is also an m-dimensional chart on M.

Any collection $\mathsf{A} = \{(U_i, \varphi_i)\}_{i\in\Lambda}$ of charts on M will be called a *partial atlas*. If their domains cover M, i.e., $M = \bigcup\limits_{i\in\Lambda} U_i$, then A is called an *atlas*. Evidently, the union of any collection of (partial) atlases is a (partial) atlas. For instance, it will also be convenient to regard each chart u as a partial atlas $\{\mathsf{u}\}$ consisting of that chart only. This will never lead to confusion.

Say that M is *locally Euclidean* if either of the following equivalent conditions holds

(LE1) every point $x \in M$ has an open neighborhood homeomorphic to some open subset of some Euclidean space (whose dimension may depend on x);

(LE2) M admits an atlas.

3.2 Differentiable Structures

Let $\mathsf{u} = (U, \varphi)$ and $\mathsf{v} = (V, \psi)$ be two *overlapping* charts on M with values in $\mathbb{R}^m$, i.e., $U \cap V \neq \varnothing$. Then the following well-defined homeomorphism between open subsets of $\mathbb{R}^m$

$$\psi \circ \varphi^{-1}\colon \varphi(U \cap V) \to \psi(U \cap V) \tag{4}$$

is called the *transition map* from u to v. Note that its inverse, $\varphi \circ \psi^{-1}\colon \psi(U \cap V) \to \varphi(U \cap V)$, is then the transition map from v to u.

Say that the transition map (4) *preserves orientation* if the determinant of the Jacobi matrix of $\psi \circ \varphi^{-1}$ is positive at each point $x \in \varphi(U \cap V)$. An atlas $\mathsf{A} = \{(U_i, \varphi_i)\}_{i\in\Lambda}$ on M is called *orientable* if all its transition maps preserve orientation.

In what follows fix once and for all some $k \in \mathbb{N} \cup \{\infty\}$. Then the charts u and v are *C^k-compatible* if either $U \cap V = \varnothing$ or the transition map $\psi \circ \varphi^{-1}$ is a C^k-diffeomorphism.

A partial atlas $\mathsf{A} = \{(U_i, \varphi_i)\}_{i \in \Lambda}$ on M is C^k if all its charts are pair-wise C^k-compatible.

Further, two partial C^k-atlases A and B are *C^k-compatible* if their union $\mathsf{A} \cup \mathsf{B}$ is still a partial C^k-atlas, i.e., the charts from A and B are pair-wise C^k-compatible as well.

Notice that the relation *"to be C^k-compatible"* on the set of all *partial atlases* on M is not an equivalence: it is reflexive and symmetric, but not transitive. However, it is easy to show that it is transitive on the set of all *atlases*:

Lemma 6 *The relation* "to be C^k-compatible" *on the set of all atlases on M is an equivalence relation. In particular, the union of any collection of pair-wise C^k-compatible C^k-atlases is still a C^k-atlas.* □

The corresponding equivalence classes of pair-wise C^k-compatible atlases on M are called *C^k-structures*. It will be convenient to say that an atlas A on M *represents* a C^k-structure $\mathfrak{A}$ if $\mathsf{A} \in \mathfrak{A}$.

It follows from Lemma 6 that the union of *all* atlases from the same C^k-structure $\mathfrak{A}$, is a *maximal atlas* of $\mathfrak{A}$, in the sense that it contains any other C^k-compatible atlas. Hence, each *C^k-structure is represented by a unique maximal atlas*, and thus we can identify C^k-structures with their maximal atlases.

However, such a description is not very useful from the "practical" point of view, since the maximal C^k-atlases are too large to be "*observable*". In many cases it is more convenient to work with atlases having as less number of charts as possible. See also Lemma 7 for another characterization of differentiable structures.

Say that a partial atlas A on M is *C^k-compatible with* $\mathfrak{A}$ or is *contained in* $\mathfrak{A}$ and write $\mathsf{A} \in \mathfrak{A}$, if either of the following equivalent conditions holds

- A is C^k-compatible with some atlas from $\mathfrak{A}$;
- A is contained in some atlas from $\mathfrak{A}$, i.e., it "extends" to some atlas on $\mathfrak{A}$;
- A is contained in the unique maximal atlas of $\mathfrak{A}$.

In particular, if $\mathsf{A} = \{\mathsf{u}\}$ consists of a single chart, then we will also write $\mathsf{u} \in \mathfrak{A}$.

A C^k-structure $\mathfrak{A}$ on M is *orientable*, if it contains an orientable atlas. Note that the latter property does not imply that *each* atlas in $\mathfrak{A}$ is also orientable.

3.3 Differentiable Maps Between Locally Euclidean Spaces

Notice that the number of *C^k-structures* on a locally Euclidean space M is very large, and in order to get some reasonable classification, one should consider those structures up to a homeomorphism being C^k-diffeomorphism between them. This leads us to discussions of smooth maps between locally Euclidean spaces endowed with C^k-structures.

Till the end of this section we assume that M and N are locally Euclidean spaces endowed with some C^k-structures $\mathfrak{A}$ and $\mathfrak{B}$ respectively.

Let $h\colon M \to N$ be a continuous map and $\mathsf{u} = (U, \varphi)$ and $\mathsf{v} = (V, \psi)$ be charts in M and N respectively such that $h(U) \subset V$. Then the induced map between open subsets of Euclidean spaces:

$$\psi \circ h \circ \varphi^{-1}\colon \varphi(U) \to \psi(V) \tag{5}$$

is called the *(coordinate) representation* of h in the charts u and v.

Say that h is C^k *(with respect to* $\mathfrak{A}$ *and* $\mathfrak{B}$), if for each $x \in M$ there exist charts $\mathsf{u} = (U, \varphi) \in \mathfrak{A}$ on M and $\mathsf{v} = (V, \psi) \in \mathfrak{B}$ on N such that

- $x \in U$, $h(x) \in V$, $h(U) \subset V$, and
- the representation $\psi \circ h \circ \varphi^{-1}\colon \varphi(U) \to \psi(V)$ of h in u and v is C^k at the point $\varphi(x)$.

In this case we will also say use the notation $h\colon (M, \mathfrak{A}) \to (N, \mathfrak{B})$ to indicate that h is a C^k-map with respect to $\mathfrak{A}$ and $\mathfrak{B}$, and denote by $C^k_{\mathfrak{A},\mathfrak{B}}(M, N)$ the set of all such maps.

A homeomorphism $h\colon M \to N$ is a C^k*-diffeomorphism (with respect to* $\mathfrak{A}$ *and* $\mathfrak{B}$*)* or *between* $(M, \mathfrak{A})$ *and* $(N, \mathfrak{B})$, if $h \in C^k_{\mathfrak{A},\mathfrak{B}}(M, N)$ and $h^{-1} \in C^k_{\mathfrak{B},\mathfrak{A}}(N, M)$.

The set of all C^k-diffeomorphisms $h\colon (M, \mathfrak{A}) \to (N, \mathfrak{B})$ will be denoted by $\mathrm{Diff}^k_{\mathfrak{A},\mathfrak{B}}(M, N)$. In particular, if $M = N$, and $\mathfrak{A} = \mathfrak{B}$, then $\mathrm{Diff}^k_{\mathfrak{A},\mathfrak{A}}(M, M)$ will also be abbreviated to $\mathrm{Diff}^k_{\mathfrak{A}}(M)$ or even to $\mathrm{Diff}^k(M)$ if the corresponding C^k-structure is assumed from the context.

The following easy lemma gives another characterization of C^k-structures.

Lemma 7 *Two* C^k*-structures* $\mathfrak{A}$ *and* $\mathfrak{B}$ *on* M *coincide if and only if* $\mathrm{id}_M \in \mathrm{Diff}^k_{\mathfrak{A},\mathfrak{B}}(M, M)$*, i.e., the identity map of* M *is a* C^k*-diffeomorphism with respect to them.* □

Example 3 Let $U \subset \mathbb{R}^m$ be an open subset. Then U admits an atlas $\mathsf{Id}_U := \{(U, \mathrm{id}_U)\}$ consisting of a single chart being just the inclusion map $\mathrm{id}_U\colon U \subset \mathbb{R}^m$. This atlas is C^k for all $k \in \mathbb{N} \cup \{\infty\}$. The corresponding C^k-structure on U will be called *canonical* and denoted by $\mathcal{I}_U$.

Also, if $V \subset \mathbb{R}^n$ is another open subset, then a map $h\colon U \to V$ belongs to $C^k_{\mathcal{I}_U,\mathcal{I}_V}(U, V)$ if and only if h is C^k as a map between open subsets of Euclidean spaces in the usual sense. □

Example 4 Let $\mathsf{A} = \{(U_i, \varphi_i)\}_{i\in\Lambda}$ be a C^k-atlas on M representing a C^k-structure $\mathfrak{A}$, and $X \subset M$ be an open subset. Then the following statements hold.

(1) X is locally Euclidean as well;
(2) $\mathsf{A}|_X := \{(X \cap U_i, \varphi_i|_{X\cap U_i}) \mid X \cap U_i \neq \varnothing, i \in \Lambda\}$ is a C^k-atlas on X;
(3) Denote by $\mathfrak{A}|_X$ the C^k-structure on X represented by $\mathsf{A}|_X$. Then the inclusion map $i\colon X \subset M$ is C^k with respect to $\mathfrak{A}|_X$ and $\mathfrak{A}$.

Moreover, let $\mathsf{v} = (V, \psi) \in \mathfrak{A}$ be an m-dimensional chart C^k-compatible with $\mathfrak{A}$. Thus $\psi(V)$ is an open subset of $\mathbb{R}^m$, and so it has a canonical C^k-structure $\mathcal{I}_{\psi(V)}$ given by the atlas $\mathsf{Id}_{\psi(V)}$. Then the corresponding homeomorphism $\psi\colon V \to \psi(V)$ is a C^k-diffeomorphism with respect to the C^k-structures $\mathfrak{A}|_V$ and $\mathcal{I}_{\psi(V)}$. □

Lemma 8 *Let $h\colon M \to N$ be a homeomorphism, $0 \le k \le \infty$, and $\mathsf{u} = (U, \varphi)$ and $\mathsf{v} = (V, \psi)$ be charts on M. Then the following statements hold.*

(1) *$h^*(\mathsf{u}) := (h(U), \varphi \circ h^{-1})$ is a chart on M with the same image $\varphi(U)$, and the representation of h in the charts u and $h^*(\mathsf{u})$ is the identity map of $\varphi(U)$:*

$$(\varphi \circ h^{-1}) \circ h \circ \varphi^{-1} = \mathrm{id}_{\varphi(U)};$$

(2) *The transition maps from u to v and from $h^*(\mathsf{u})$ to $h^*(\mathsf{v})$ coincide, i.e.,*

$$(\psi \circ h^{-1}) \circ (\varphi \circ h^{-1})^{-1} = \psi \circ \varphi^{-1};$$

(3) *Let $\mathsf{A} = \{(U_i, \varphi_i)\}_{i\in\Lambda}$ be a C^k-atlas on M representing a C^k-structure $\mathfrak{A}$. Then*

$$h^*\mathsf{A} = \{(h(U_i), \varphi_i \circ h^{-1})\}_{i\in\Lambda}$$

is an atlas on N with the same transition maps between the charts with the corresponding indices. In particular, the following statements hold.

(a) *$h^*\mathsf{A}$ is of class C^k as well, and we will denote its C^k-structure by $h^*\mathfrak{A}$;*
(b) *A is a maximal C^k-atlas iff so is $h^*\mathsf{A}$;*
(c) *If $k \ge 1$, then h is a C^k-*diffeomorphism *with respect to $\mathfrak{A}$ and $h^*\mathfrak{A}$, i.e., $h \in \mathrm{Diff}^k_{\mathfrak{A}, h^*\mathfrak{A}}(M, M)$.* □

Thus, due to Lemma 8(3b), for each $k \in \mathbb{N} \cup \{\infty\}$ the group $\mathcal{H}(M)$ of homeomorphisms of M naturally acts on the set of C^k-structures, and these homeomorphisms are diffeomorphisms between the corresponding C^k-structures. Therefore the *right* formulation of the problem of classification of C^k-structures on M reduces to the following one:

Problem *Describe the orbits of the action of the group $\mathcal{H}(M)$ on the set of C^k-structures on M.* □

In particular, the statement that *each manifold M of dimension ≤ 3 has a unique* (*up to a diffeomorphism*) *C^k-structure* means that the above action is transitive. In next section we will present an elementary proof of that fact for $\dim M = 1$. Such a statement is usually referred as a classical result, though the authors does not find its proof in the available literature. In fact, we prove some technical statement which will be useful in this paper for studying differentiable structures on non-Hausdorff 1-manifolds.

Let us also mention two simple statements which will be used several times.

Lemma 9 *Let X be an open subset of $\mathbb{R}^n$ for some $n \ge 0$, and $\mathfrak{A}$ and $\mathfrak{B}$ be two C^k-structures on X represented by atlases $\{(X, \varphi)\} \in \mathfrak{A}$ and $\{(X, \psi)\} \in \mathfrak{B}$ consisting of single charts and such that $\varphi(X) = \psi(X) \subset \mathbb{R}^n$. Then the homeomorphism*

$$\psi^{-1} \circ \varphi\colon X \xrightarrow{\varphi} \varphi(X) = \psi(X) \xrightarrow{\psi^{-1}} X$$

is a C^k-diffeomorphism between $(X, \mathfrak{A})$ and $(X, \mathfrak{B})$, i.e., $\psi^{-1} \circ \varphi \in \mathrm{Diff}^k_{,}(\mathfrak{A}, \mathfrak{B})XX$. □

Proof Consider the following commutative diagram:

$$\begin{array}{ccc} X & \xrightarrow{\ \psi^{-1}\circ\varphi\ } & X \\ {\scriptstyle\varphi}\downarrow & & \downarrow{\scriptstyle\psi} \\ \varphi(X) & \overset{\mathrm{id}}{=\!=\!=} & \psi(X) \end{array}$$

It means that the representation of $\psi^{-1} \circ \varphi$ in the charts u and v is the identity map id: $\varphi(X) = \psi(X)$ which is a diffeomorphism of class C^k. Hence $\psi^{-1} \circ \varphi$ is a C^k-diffeomorphism. □

Lemma 10 *Let M and N be two locally Euclidean spaces endowed with C^k-structures $\mathfrak{A}$ and $\mathfrak{B}$ respectively. Suppose there exists a C^k-diffeomorphism $h \in \mathrm{Diff}^k_{\mathfrak{A},\mathfrak{B}}(M, N)$. Then we have the following commutative diagram of bijections:*

$$\begin{array}{ccccc} & & \mathrm{Diff}^k_{\mathfrak{A},\mathfrak{B}}(M,N) & & \\ & {\scriptstyle g\mapsto h\circ g}\nearrow & & \searrow{\scriptstyle k\mapsto k\circ h^{-1}} & \\ \mathrm{Diff}^k_{\mathfrak{A}}(M) & & \xrightarrow{\ g\mapsto h\circ g\circ h^{-1}\ } & & \mathrm{Diff}^k_{\mathfrak{B}}(N) \end{array}$$

in which the lower horizontal arrow is an isomorphism of groups. In particular, if the groups $\mathrm{Diff}^k_{\mathfrak{A}}(M)$ and $\mathrm{Diff}^k_{\mathfrak{B}}(N)$ are not isomorphic, then $(M, \mathfrak{A})$ and $(N, \mathfrak{B})$ are not C^k-diffeomorphic. □

The inverse statement, i.e., that every isomorphism $\alpha\colon \mathrm{Diff}^k_{\mathfrak{A}}(M) \to \mathrm{Diff}^k_{\mathfrak{B}}(N)$ is induced by some diffeomorphism between M and N, is called a *reconstruction theorem*. It holds for Hausdorff second countable manifolds M and N, see [1, 20, 29, 30, 40] for precise statements and references therein.

4 Uniqueness of C^k-Structures on $\mathbb{R}$ Up to a Diffeomorphism

In this section M is a one-dimensional locally Euclidean space and $k \in \mathbb{N} \cup \{\infty\}$. We will describe a way of joining charts on M. This technique is mentioned in [15].

Definition 1.1 Let $\mathsf{u} = (U, \varphi)$ and $\mathsf{v} = (V, \psi)$ be two charts on M. Say that u and v are C^k-*joinable* if they are C^k-compatible and

$$\varphi(U) = (a; c), \qquad \varphi(U \cap V) = \psi(U \cap V) = (b; c), \qquad \psi(V) = (b; d)$$

for some $a < b < c < d \in \mathbb{R}$.

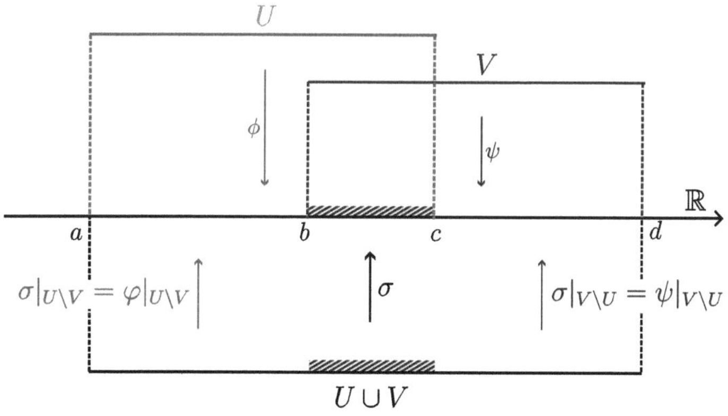

Fig. 3 C^k-joinable charts

In this case, see Fig. 3, every open embedding $\sigma : U \cup V \to \mathbb{R}$, regarded as a chart $\mathsf{w} = (U \cup V, \sigma)$ on M, will be called a C^k*-join* of u and v if the following conditions hold:

(1) $\sigma(U) = \varphi(U) = (a; c)$ and $\sigma(V) = \psi(V) = (b; d)$;
(2) w is C^k-compatible with u and v;
(3) $\sigma|_{U\setminus V} = \varphi|_{U\setminus V} : U \setminus V \to (a; c]$ and $\sigma|_{V\setminus U} = \psi|_{V\setminus U} : V \setminus U \to [b; d)$.

The first condition implies that $\sigma(U \cap V) = (c; b)$, while the latter condition can be formulated so that σ *differs from* φ *and* ψ *only on the intersection* $U \cap V$, or equivalently, that the transition maps $\sigma \circ \varphi^{-1}$ and $\sigma \circ \psi^{-1}$ are identity on $U \setminus V$ and $V \setminus U$ respectively.

Our aim is to establish the following theorem

Theorem 1.1 *(1) Every pair of* C^k*-joinable charts* u *and* v *has a* C^k*-join.*
(2) Let A *be a* C^k*-atlas on* M *having a pair of* C^k*-joinable charts* u *and* v *such that for every other chart* $\mathsf{y} = (Y, \xi)$ *from* A *we have that*

$$Y \cap U \cap V = \varnothing.$$

Then for any C^k*-join* w *of* u *and* v *the following collection of charts*

$$\mathsf{B} = \big(\mathsf{A} \setminus \{\mathsf{u}, \mathsf{v}\}\big) \cup \{\mathsf{w}\},$$

obtained by replacing u *and* v *with* w*, is still a* C^k*-atlas on* M *being also* C^k*-compatible with* A.

Proof First we establish the following simple lemma.

Lemma 11 *Let* $b < c \in \mathbb{R}$ *and* $g : (b; c) \to (b; c)$ *be a* C^k*-diffeomorphism preserving orientation. Then there exist* $\varepsilon > 0$ *and a diffeomorphism* $p : (b; c) \to (b; c)$ *such that*

$$p(x) = x \textit{ for } x \in (b; b+\varepsilon], \qquad p(x) = g(x) \textit{ for } x \in [c-\varepsilon; c).$$

Proof It suffices to find $\varepsilon > 0$ and a strictly positive function $\gamma\colon [b; c) \to (0; +\infty)$ such that

$$\gamma(x) = 1 \text{ for } x \in [b; b+\varepsilon], \quad \gamma(x) = g'(x) \text{ for } x \in [c-\varepsilon; c), \quad \int_b^c \gamma(s)ds = c - b,$$

and then put $p(x) = b + \int_b^x \gamma(s)ds$.

As g is a preserving orientation homeomorphism of $(b; c)$, we have that

$$b = \lim_{x\to b+0} g(x) < \lim_{x\to c-0} g(x) = c.$$

Then one can easily find $\varepsilon > 0$ and two C^∞ functions $\alpha, \beta\colon [b; c] \to [0; +\infty)$ such that

$$\alpha = 1 \text{ on } [b; b+\varepsilon], \qquad \alpha = 0 \text{ on } [b+2\varepsilon; c-2\varepsilon], \qquad \alpha = g' \text{ on } [c-\varepsilon; c],$$
$$A := c - b - \int_b^c \alpha(s)ds > 0, \qquad \beta^{-1}(0) = [b; b+\varepsilon] \cup [c-\varepsilon; c].$$

Denote $B = \int_b^c \beta(s)ds$. Then the function $\gamma = \alpha + (A/B)\beta$ has the above properties.

Now we can prove Theorem 1.1.

(1) Let $\mathsf{u} = (U, \varphi)$ and $\mathsf{v} = (V, \psi)$ be two C^k-joinable charts on M. Not loosing generality, assume that

$$\varphi(U) = (0; 2), \qquad \varphi(U \cap V) = \psi(U \cap V) = (1; 2), \qquad \psi(V) = (1; 3).$$

Denote $W = U \cup V$. We need to construct a C^k-join $\mathsf{w} = (W, \sigma)$ for these charts.

Let $g = \psi \circ \varphi^{-1}\colon (1; 2) \to (1; 2)$ be the transition map from u to v. Then by Lemma 11 one can find $\varepsilon > 0$ and a C^k-diffeomorphism $p\colon (0; 2) \to (0; 2)$ such that

$$p(x) = x \text{ for } x \in (0; 1+\varepsilon), \tag{6}$$
$$p(x) = g(x) \text{ for } x \in (2-\varepsilon; 2). \tag{7}$$

Consider the homeomorphism $\bar{\varphi} := p \circ \varphi\colon U \to (0; 2)$ regarded as a chart $\bar{\mathsf{u}} = (U, \bar{\varphi})$. Then by (6), $\bar{\varphi} = \varphi$ on $(0; 1+\varepsilon)$, whence

$$\bar{\varphi}(U \cap V) = \varphi(U \cap V) = \psi(U \cap V) = (1; 2).$$

Moreover, by (7), the transition map from v to $\bar{\mathsf{u}}$, being a C^k-diffeomorphism

$$q := \bar{\varphi} \circ \psi^{-1} = p \circ \varphi \circ \psi^{-1} = p \circ g^{-1}\colon (1; 2) \to (1; 2)$$

is fixed on $(2-\varepsilon'; 2)$ for a sufficiently small $\varepsilon' > 0$. Hence q extends by the identity on $(2; 3)$ to a self-diffeomorphism of $(1; 3)$, which we will denote by the same letter q.

Since $\bar{\varphi}|_{U\cap V} = q \circ \psi|_{U\cap V}$, we get a well-defined homeomorphism

$$\sigma : W = U \cup V \to (0; 3), \qquad \sigma(x) = \begin{cases} \bar{\varphi}(x), & x \in U, \\ q \circ \psi(x), & x \in V, \end{cases}$$

i.e., the chart $\mathsf{w} = (W, \sigma)$.

Note that p and q are the transition maps from u and v to w respectively:

$$\begin{aligned} \sigma \circ \varphi^{-1} &= \bar{\varphi} \circ \varphi^{-1} = p \circ \varphi \circ \varphi^{-1} = p\colon (0; 2) \to (0; 2), \\ \sigma \circ \psi^{-1} &= q \circ \psi \circ \psi^{-1} = q\colon (1; 3) \to (1; 3). \end{aligned}$$

Since they are C^k-diffeomorphisms, p is fixed on $U \setminus V$, and q is fixed on $V \setminus U$, we see that w is a C^k-join of u and v.

(2) Let $\mathsf{u} = (U, \varphi)$ and $\mathsf{v} = (V, \psi)$ be two C^k-joinable charts in the atlas A such that for every other chart $\mathsf{y} = (Y, \xi)$ from A we have that $Y \cap U \cap V = \varnothing$. Let also w be any C^k-join of u and v. We need to prove that

$$\mathsf{B} = \big(\mathsf{A} \setminus \{\mathsf{u}, \mathsf{v}\}\big) \cup \{\mathsf{w}\}$$

is a C^k-atlas on M being also C^k-compatible with A.

Since $W = U \cup V$, the domains of charts of B cover M.

Let us show that w *is* C^k*-compatible with all other charts* $\mathsf{y} = (Y, \xi) \in \mathsf{B}$ *distinct from* w, i.e., the transition map $\sigma \circ \xi^{-1}\colon \xi(Y \cap W) \to \sigma(Y \cap W)$ is a C^k-diffeomorphism.

Indeed, since $Y \cap U \cap V = \varnothing$, we have that $Y \cap W$ is a union of disjoint open subsets $Y \cap U$ and $Y \cap V$, i.e., $Y \cap W = (Y \cap U) \sqcup (Y \cap V)$. Since σ coincides with φ on $U \setminus V \supset Y \cap U$ and with ψ on $V \setminus U \supset Y \cap V$, we see that

$$\sigma \circ \xi^{-1}|_{\xi(Y\cap U)} = \varphi \circ \xi^{-1}|_{\xi(Y\cap U)}, \qquad \sigma \circ \xi^{-1}|_{\xi(Y\cap V)} = \psi \circ \xi^{-1}|_{\xi(Y\cap V)}. \tag{8}$$

But A is a C^k-atlas, i.e., y is C^k-compatible with u and v, whence both restrictions (8) are C^k, and therefore $\sigma \circ \xi^{-1}$ is C^k as well. □

4.1 Chain-Like Atlases

Let $\mathsf{B} = \{\mathsf{v}_j = (V_j, \psi_j)\}_{j\in\mathbb{Z}}$ be a countable C^k-atlas on the real line $\mathbb{R}$. Say that B is *chain-like* if for each $j \in \mathbb{Z}$ the following conditions hold:

(G1) the charts v_j and v_{j+1} are C^k-joinable;
(G2) $V_i \cap V_j \cap V_{j+1} = \varnothing$ for all $i \neq j, j+1$.

Equivalently, suppose that $\psi_j(V_j) = (a_j; b_j)$ for some $a_j < b_j \in \mathbb{R} \cup \{\pm\infty\}$, $j \in \mathbb{Z}$. Then B is *chain-like* if for every $j \in \mathbb{Z}$ we have that

- all a_j, b_j are finite,
- $a_j < b_{j-1} < a_{j+1} < b_j$,
- $\psi_j(V_j \cap V_{j+1}) = \psi_{j+1}(V_j \cap V_{j+1}) = (a_{j+1}; b_j)$.

Notice that for any pairs of strictly increasing two-sided sequences $\{x_i\}_{i\in\mathbb{Z}}$ and $\{y_i\}_{i\in\mathbb{Z}}$ in $\mathbb{R}$ there exists a C^∞-diffeomorphism $q\colon \mathbb{R} \to \mathbb{R}$ such that $q(x_i) = y_i$ for $i \in \mathbb{Z}$. Hence one can always assume that for such a chain-like atlas we have that

$$\lim_{j\to-\infty} a_j = \lim_{j\to-\infty} b_j = -\infty, \qquad \lim_{j\to+\infty} a_j = \lim_{j\to+\infty} b_j = +\infty, \tag{9}$$

which is equivalent to the assumption that

(G3) $\bigcup\limits_{j\in\mathbb{Z}} \psi_j(V_j) = \mathbb{R}$.

Theorem 3 (1) *For every C^k-atlas $\mathsf{A} = \{(U_i, \varphi_i)\}_{i\in\Lambda}$ on $\mathbb{R}$ there exists a chain-like C^k-atlas B being C^k-compatible with A.*
(2) *Let B be a chain-like C^k-atlas on $\mathbb{R}$. Then there exists a homeomorphism $\sigma : \mathbb{R} \to \mathbb{R}$ such that the chart $(\mathbb{R}, \sigma)$ is C^k-compatible with B.* □

Proof (1) Since $\mathbb{R}$ is one-dimensional and paracompact, and the domains of A constitute an open cover of $\mathbb{R}$, one can find a countable $\{V_j\}_{j\in\mathbb{Z}}$ cover of $\mathbb{R}$ such that for every $j \in \mathbb{Z}$

(a) $V_j = (a_j; b_j)$ for some $a_j < b_j \in \mathbb{R}$ such that $a_j < b_{j-1} < a_{j+1} < b_j$, i.e., V_j intersects only V_{j-1} and V_{j+1} and those intersections are proper subsets of V_j, so in particular,
$$V_i \cap V_j \cap V_{j+1} = \varnothing, \qquad i \neq j, j+1; \tag{10}$$

(b) there exists a chart $\mathsf{u}_j = (U_j, \varphi_j) \in \mathsf{A}$ with $V_j \subset U_j$;

Note that for any C^k-embedding $p_j\colon \varphi_j(V_j) \to \mathbb{R}$, the chart $\mathsf{v}_j := (V_j, \psi_j := p_j \circ \varphi_j|_{V_j})$ is still C^k-compatible with A. In particular, one can choose p_j so that

$$\psi_j(V_j) = (a_j; b_j), \quad \psi_j(V_{j-1} \cap V_j) = (a_j; b_{j-1}) \quad \psi_j(V_j \cap V_{j+1}) = (a_{j+1}; b_j), \tag{11}$$

which will imply that v_j and v_{j+1} are C^k-compatible for all $j \in \mathbb{Z}$.

Indeed, note $\varphi_j(V_j) = (c_j; d_j)$ for some $c_j < d_j \in \mathbb{R}$. Take $p_j\colon (c_j; d_j) \to (a_j; b_j)$ to be any C^k-diffeomorphism such that $p_j \circ \varphi_j(b_{j-1}) = b_{j-1}$ and $p_j \circ \varphi_j(a_{j+1}) = a_{j+1}$. Then the composite map

$$\psi_j := p_j \circ \varphi_j|_{V_j} \colon V_j = (a_j; b_j) \xrightarrow{\varphi_j} (c_j; d_j) \xrightarrow{p_j} (a_j; b_j) \subset \mathbb{R}$$

has the desired property (11).

Notice that (11) and (10) mean respectively conditions (G1) and (G2), so $\mathsf{B} := \{(V_j, \psi_j)\}_{j\in\mathbb{Z}}$ is a chain-like C^k-atlas on $\mathbb{R}$.

(2) Suppose now that $\mathbb{R}$ has a chain-like atlas $\mathsf{B} = \{\mathsf{v}_j = (V_j, \psi_j)\}_{j\in\mathbb{Z}}$. As noted above, one can assume that condition (9) is also satisfied.

Then by Theorem 1.1 there exists a C^k-join $\mathsf{w}_{0,1} = (V_0 \cup V_1, \sigma_{0,1})$ of v_0 and v_1, such that the collection of charts

$$\mathsf{B}_{0,1} := \big(\mathsf{B} \setminus \{\mathsf{v}_0, \mathsf{v}_1\}\big) \cup \{\mathsf{w}_{0,1}\}$$

is a C^k-atlas being C^k-compatible with B. By definition, $\mathsf{w}_{0,1}$ differs from ψ_0 and ψ_1 only on the intersection $V_0 \cap V_1$ which is disjoint from other chars from B. This easily implies that $\mathsf{B}_{0,1}$ is still chain-like.

In particular, the charts v_{-1} and $\mathsf{w}_{0,1}$ are C^k-joinable, and again by Theorem 1.1 for any of their C^k-join

$$\mathsf{w}_{-1,1} = \big(V_{-1} \cup W_{0,1} = V_{-1} \cup V_0 \cup V_1,\ \sigma_{-1,1}\big)$$

the collection of charts

$$\mathsf{B}_{-1,1} := \big(\mathsf{B} \setminus \{\mathsf{v}_{-1}, \mathsf{w}_{0,1}\}\big) \cup \{\mathsf{w}_{-1,1}\} = \big(\mathsf{B} \setminus \{\mathsf{v}_{-1}, \mathsf{v}_0, \mathsf{v}_1\}\big) \cup \{\mathsf{w}_{-1,1}\}$$

is a chain-like C^k-atlas being C^k-compatible with $\mathsf{B}_{0,1}$ and therefore with B. Again notice that $\sigma_{-1,1}$ differs from with ψ_{-1} and $\sigma_{0,1}$ only on the intersection

$$V_{-1} \cap (V_0 \cap V_1) = V_{-1} \cap V_0.$$

In particular, $\sigma_{-1,1} = \psi_0$ on the compact segment

$$A_1 := V_0 \setminus (V_{-1} \cup V_1) = \psi_0^{-1}([b_{-1}, a_1]).$$

Thus, by induction for each $n \geq 1$ we can construct a sequence of chain-like C^k-atlases

$$\mathsf{B}_{-n,n} := \big(\mathsf{B} \setminus \{\mathsf{v}_j\}_{j=-n}^{n}\big) \cup \{\mathsf{w}_{-n,n}\},$$

where $\mathsf{w}_{-n,n} = \big(\bigcup_{j=-n}^{n} V_j, \sigma_{-n,n}\big)$ is a chart such that $\sigma_{-n,n}\big(\bigcup_{j=-n}^{n} V_j\big) = (a_{-n}; b_n)$ and $\sigma_{-n,n} = \sigma_{-n+1,n-1}$ on the compact segment

$$A_n := \big(\bigcup_{j=-n}^{n} V_j\big) \setminus \big(V_{-n} \cup V_n\big).$$

Since $A_n \subset \operatorname{Int} A_{n+1}$ and $\mathbb{R} = \bigcup_{n=0}^{+\infty} A_n$, for each $x \in \mathbb{R}$ there exists $n > 0$ such that

$$\sigma_{-i,i}(x) = \sigma_{-j,j}(x) \text{ for all } i, j \geq n.$$

Then due to (9), we get a well-defined homeomorphism $\sigma\colon \mathbb{R} \to \mathbb{R}$, $\sigma(x) = \lim_{n\to\infty} \sigma_{-n,n}(x)$, which is C^k-compatible with all the above atlases and, in particular, with B. □

Corollary 1.1 *Every C^k-structure $\mathfrak{B}$ on $\mathbb{R}$ is C^k-diffeomorphic with the canonical one given by the atlas $\mathsf{Id}_{\mathbb{R}} = \{(\mathbb{R}, \mathrm{id}_{\mathbb{R}})\}$ with a single chart.*

Proof By Theorem 3, $\mathfrak{B}$ is represented by an atlas $\mathsf{B} = \{(\mathbb{R}, \psi)\}$ consisting of a single chart given by some homeomorphism $\psi\colon \mathbb{R} \to \mathbb{R}$. On the other hand, the canonical C^k-structure $\mathcal{I}_{\mathbb{R}}$ is also given by the atlas with a single chart whose coordinate homeomorphism $\mathrm{id}\colon \mathbb{R} \to \mathbb{R}$ has the same image $\mathbb{R}$ as ψ. Then by Lemma 9, the map $\psi = \psi \circ \mathrm{id}_{\mathbb{R}}^{-1}\colon \mathbb{R} \to \mathbb{R}$ is a C^k-diffeomorphism between $(\mathbb{R}, \mathfrak{B})$ and $(\mathbb{R}, \mathcal{I}_{\mathbb{R}})$. □

Corollary 1 *Let M be a one-dimensional locally Euclidean topological space, A a C^k-atlas on M, and $U \subset M$ be an open subset homeomorphic with $\mathbb{R}$. Then there exists a homeomorphism $\varphi\colon U \to \mathbb{R}$ being C^k-compatible with A. In other words, $\mathsf{A} \cup \{(U, \varphi)\}$ is a C^k-atlas on M being C^k-compatible with A.* □

We finish this section with the following well-known statement which could be proved in using the approach of similar to Theorem 3.

Theorem 1.2 *Let M be a connected Hausdorff one-dimensional locally Euclidean topological space. Then for any two C^k-structures $\mathfrak{A}$ and $\mathfrak{B}$ on M there exists a C^k-diffeomorphism $h \in \mathrm{Diff}^k_{\mathfrak{A},\mathfrak{B}}(M, M)$. In other words, the action of the group of homeomorphisms $\mathcal{H}(M)$ on the set of all C^k-structures on M is transitive.*

5 Differentiable Structures on the Line with Double Origin

It this section we will give a classification of C^k-structures on the line with two origins $\mathbb{L} = \mathbb{R} \sqcup \{\bar{0}\}$ for all $k = 1, 2, \ldots, \infty$. It will be convenient to describe at first the structure of the groups $\mathcal{H}(\mathbb{L})$ of all homeomorphisms of $\mathbb{L}$, and then compare it with the subgroups of diffeomorphisms of distinct differentiable structures. Denote $U := \mathbb{R} = \mathbb{L} \setminus \bar{0}$ and $V := \mathbb{L} \setminus 0$.

5.1 *Homeomorphisms of* $\mathbb{R}$ *and* $\mathbb{L}$

Consider the diffeomorphism $q\colon \mathbb{R} \to \mathbb{R}$, $q(x) = -x$, reversing orientation and having order 2. Evidently, $q \in \mathcal{E}^{\infty,-}(\mathbb{R}, 0)$.

Recall that $\mathcal{W}^k(\mathbb{R}, 0)$ denotes the group of homeomorphisms h of $\mathbb{R}$ such that $h(0) = 0$ and the restriction $h|_{\mathbb{R}\setminus 0}\colon \mathbb{R} \setminus 0 \to \mathbb{R} \setminus 0$ is a C^k-diffeomorphism of $\mathbb{R} \setminus 0$, $\mathrm{Diff}^k(\mathbb{R}, 0)$ is the subgroup of $\mathcal{W}^k(\mathbb{R}, 0)$ consisting of C^k-diffeomorphisms of $\mathbb{R}$, and $\mathcal{E}^k(\mathbb{R}, 0)$ is the subgroup of $\mathrm{Diff}^k(\mathbb{R}, 0)$ consisting of diffeomorphisms h such that $h^{(i)}(0) = 0$ for $i = 2, 3, \ldots, k$. We also put upper index "+" or "−" to indicate homeomorphisms respectively preserving or reversing the orientation of $\mathbb{R}$.

Let $\mathcal{G}$ be one of the groups $\mathcal{H}^+(\mathbb{R}, 0)$, $\mathcal{W}^{k,+}(\mathbb{R}, 0)$, $\mathrm{Diff}^{k,+}(\mathbb{R}, 0)$ or $\mathcal{E}^{k,+}(\mathbb{R}, 0)$ defined in Sect. 1.1, and $\mathcal{G}^+ = \mathcal{G} \cap \mathcal{H}^+(\mathbb{R}, 0)$.

Then it is easy to see that $q\,\mathcal{G}^+ q^{-1} = \mathcal{G}^+$, so the conjugation by q is an automorphism of $\mathcal{G}^+$ of order 2. Hence one can define the semidirect product $\mathcal{G}^+ \rtimes \mathbb{Z}_2$ corresponding to that automorphism. Thus, by definition, $\mathcal{G}^+ \rtimes \mathbb{Z}_2$ is a Cartesian product $\mathcal{G}^+ \times \mathbb{Z}_2$ with the following multiplication $(h, \delta)(g, \varepsilon) = (h \circ q^{\delta} \circ g \circ q^{-\delta}, \delta\varepsilon)$. In particular, the inclusion $\mathcal{G}^+ \subset \mathcal{H}^+(\mathbb{R}, 0)$ allows to consider $\mathcal{G}^+ \rtimes \mathbb{Z}_2$ as a subgroup of $\mathcal{H}^+(\mathbb{R}, 0) \rtimes \mathbb{Z}_2$.

The following two lemmas are straightforward, see notations in Sect. 2.1, and we leave them for the reader. They describe certain algebraic properties of the groups of homeomorphisms of $\mathbb{R}$ and $\mathbb{L}$ and also establish relationships between them.

Lemma 12 *The map $\eta\colon \mathcal{H}^+(\mathbb{R}, 0) \rtimes \mathbb{Z}_2 \to \mathcal{H}(\mathbb{R}, 0)$, $\eta(h, \delta) = h \circ q^{\delta}$, is an isomorphism making commutative the following diagram:*

$$
\begin{array}{ccccc}
\mathcal{H}^+(\mathbb{R}, 0) & \xhookrightarrow{g \mapsto (g,0)} & \mathcal{H}^+(\mathbb{R}, 0) \rtimes_q \mathbb{Z}_2 & \xrightarrow{(g,a) \mapsto a} & \mathbb{Z}_2 \\
\Vert & & \eta \downarrow \cong & & \Vert \\
\mathcal{H}^+(\mathbb{R}, 0) & \hookrightarrow & \mathcal{H}(\mathbb{R}, 0) & \xrightarrow{s} & \langle q \rangle \cong \mathcal{H}(\mathbb{R}, 0)/\mathcal{H}^+(\mathbb{R}, 0)
\end{array}
$$

and isomorphically mapping $\mathcal{G}^+ \times \mathbb{Z}_2$ onto $\mathcal{G}$. □

Lemma 13 (1) *Let $\zeta : \mathbb{L} \to \mathbb{L}$ be the (unique) map exchanging 0 and $\bar{0}$ and fixed on the complement $\mathbb{L} \setminus \{0, \bar{0}\}$. Then*

(a) *ζ is a homeomorphism of $\mathbb{L}$ and $\zeta^2 = \mathrm{id}_{\mathbb{L}}$, so it generates the subgroup $\langle \zeta \rangle \cong \mathbb{Z}_2$;*

(b) *ζ commutes with each $h \in \mathcal{H}(\mathbb{L})$ and $\langle \zeta \rangle$ coincides with the center of $\mathcal{H}(\mathbb{L})$.* □

(2) *$\mathcal{H}^{\mathrm{fix}}(\mathbb{L})$ is the kernel of the "restriction to $\{0, \bar{0}\}$" homomorphism*

$$\mathsf{p}\colon \mathcal{H}(\mathbb{L}) \to \mathrm{Perm}(\{0, \bar{0}\}) = \mathbb{Z}_2, \qquad \mathsf{p}(h) = h|_{\{0,\bar{0}\}}$$

to the group of permutations of $\{0, \bar{0}\}$. Moreover, the "restriction to U" map

$$\mathsf{r}_U : \mathcal{H}^{\mathrm{fix}}(\mathbb{L}) \to \mathcal{H}(\mathbb{R}, 0), \qquad \mathsf{r}_U(h) = h|_U$$

is an isomorphism of groups. The inverse map $\mathsf{r}_U^{-1} : \mathcal{H}(\mathbb{R}, 0) \to \mathcal{H}^{\mathrm{fix}}(\mathbb{L})$ associates to each $h \in \mathcal{H}(\mathbb{R}, 0)$ its unique extension to a homeomorphism of all $\mathbb{L}$ fixing $\bar{0}$.

(3) *The normal subgroups* $\mathcal{H}^{\mathrm{fix}}(\mathbb{L})$ *and* $\langle\zeta\rangle$ *generate* $\mathcal{H}(\mathbb{L})$*, and* $\mathcal{H}^{\mathrm{fix}}(\mathbb{L}) \cap \langle\zeta\rangle = \{\mathrm{id}_{\mathbb{L}}\}$*, whence* $\mathcal{H}(\mathbb{L})$ *splits into the direct product* $\mathcal{H}^{\mathrm{fix}}(\mathbb{L}) \times \langle\zeta\rangle$*. More explicitly, the map*

$$\alpha\colon \mathcal{H}^{\mathrm{fix}}(\mathbb{L}) \times \langle\zeta\rangle \equiv \mathcal{H}(\mathbb{R}, 0) \times \mathbb{Z}_2 \to \mathcal{H}(\mathbb{L}), \qquad \alpha(g, \varepsilon) = g \circ \xi^{\varepsilon}$$

is an isomorphism of groups making commutative the following diagram:

$$\begin{array}{ccccc}
\mathcal{H}^{\mathrm{fix}}(\mathbb{L}) & \hookrightarrow & \mathcal{H}(\mathbb{L}) & \xrightarrow{\mathsf{p}\colon h \mapsto h|_{\{0,\bar{0}\}}} & \mathrm{Perm}(\{0, \bar{0}\}) \\
{\scriptstyle \mathsf{r}_U\colon h \mapsto h|_U} \downarrow \cong & & \alpha \uparrow \cong & & \| \\
\mathcal{H}(\mathbb{R}, 0) & \xrightarrow{g \mapsto (g,0)} & \mathcal{H}(\mathbb{R}, 0) \times \mathbb{Z}_2 & \xrightarrow{(g,\varepsilon) \mapsto \varepsilon} & \mathbb{Z}_2
\end{array}$$

(4) *Let* $q \in \mathcal{H}(\mathbb{R}, 0)$ *be any homeomorphism such that* $q^2 = \mathrm{id}_{\mathbb{R}}$*, and* $\hat{q} = \mathsf{r}_U^{-1}(q) \in \mathcal{H}^{\mathrm{fix}}(\mathbb{L})$ *be it unique extension to all of* $\mathbb{L}$ *fixing* $\bar{0}$*. Then we have an isomorphism*

$$\big(\mathcal{H}^{+}(\mathbb{R}, 0) \rtimes_q \mathbb{Z}_2\big) \times \mathbb{Z}_2 \cong \mathcal{H}(\mathbb{L}), \qquad (g, a, \varepsilon) \mapsto g \circ \hat{q}^a \circ \zeta^{\varepsilon}.$$

5.2 *Group* $\mathbf{Diff}^{k}_{\mathbf{mon}}(\mathbb{R} \setminus \mathbf{0})$

From now on fix some $k \in \mathbb{N} \cup \{\infty\}$, and let $\mathrm{Diff}^k_{\mathrm{mon}}(\mathbb{R} \setminus 0)$ be the group of *monotone* C^k-diffeomorphisms of $\mathbb{R} \setminus 0$. Thus each $h \in \mathrm{Diff}^k_{\mathrm{mon}}(\mathbb{R} \setminus 0)$ is either increasing or decreasing, so it

- either leaves invariant each of intervals $(-\infty, 0)$ and $(0, +\infty)$ and preserves their orientation,
- or exchanges those intervals and reverses their orientations.

Example 5 The diffeomorphism $h\colon \mathbb{R} \setminus 0 \to \mathbb{R} \setminus 0$, $h(x) = 1/x$, is *not monotone* and therefore does not belong to $\mathrm{Diff}^k_{\mathrm{mon}}(\mathbb{R} \setminus 0)$. □

Let $h \in \mathcal{W}^k(\mathbb{R}, 0)$, so it is a homeomorphism of $\mathbb{R}$ fixing 0 and being a C^k-diffeomorphism of $\mathbb{R} \setminus 0$. In particular, its restriction $h|_{\mathbb{R}\setminus 0}\colon \mathbb{R} \setminus 0 \to \mathbb{R} \setminus 0$ is a monotone C^k-diffeomorphism.

Lemma 14 *The map* $\gamma\colon \mathcal{W}^k(\mathbb{R}, 0) \to \mathrm{Diff}^k_{\mathrm{mon}}(\mathbb{R} \setminus 0)$*,* $\gamma(h) = h|_{\mathbb{R}\setminus 0}$*, is an isomorphism of groups.* □

Proof Evidently, γ is a monomorphism, and we need to show that γ is surjective. Let $g \in \mathrm{Diff}^k_{\mathrm{mon}}(\mathbb{R} \setminus 0)$. Since g is monotone, $\lim\limits_{x\to 0} g(x) = 0$, whence g extends to a homeomorphism $\hat{g}\colon \mathbb{R} \to \mathbb{R}$, defined by $\hat{g} = g$ on $\mathbb{R} \setminus 0$ and $\hat{g}(0) = 0$. It follows, that $\hat{g} \in \mathcal{W}^k(\mathbb{R}, 0)$ and $\gamma(\hat{g}) = g$, i.e., γ is surjective as well. □

For each $g \in \mathrm{Diff}^k_{\mathrm{mon}}(\mathbb{R} \setminus 0)$ the homeomorphism $\gamma^{-1}(g)$ will be called the 0-*extension* of g. Note also that we have the following inclusion

$$\mathrm{Diff}^k(\mathbb{R}, 0) \subset \mathcal{W}^k(\mathbb{R}, 0) \overset{\gamma^{-1}}{\equiv} \mathrm{Diff}^k_{\mathrm{mon}}(\mathbb{R} \setminus 0).$$

Thus, one can regard $\mathrm{Diff}^k(\mathbb{R}, 0)$ as a subgroup of $\mathrm{Diff}^k_{\mathrm{mon}}(\mathbb{R} \setminus 0)$. It is easy to check that $\mathrm{Diff}^k(\mathbb{R}, 0)$ is not normal in $\mathrm{Diff}^k_{\mathrm{mon}}(\mathbb{R} \setminus 0)$.

Lemma 15 *The subgroup* $\mathrm{Diff}^k_{\mathrm{nb}}(\mathbb{R}, 0) \subset \mathcal{W}^k(\mathbb{R}, 0)$ *consisting of diffeomorphisms of* $\mathbb{R}$ *fixed near* 0 *is a normal.* □

Proof Let $h \in \mathrm{Diff}^k_{\mathrm{nb}}(\mathbb{R}, 0)$ and $g \in \mathcal{W}^k(\mathbb{R}, 0)$. Then $g \circ h \circ g^{-1}$ is fixed near $0 \in \mathbb{R}$ and its restriction to $\mathbb{R} \setminus 0$ is a C^k-diffeomorphism. Hence $g \circ h \circ g^{-1}$ is a C^k-diffeomorphism of $\mathbb{R}$ fixed near 0, i.e., it belongs to $\mathrm{Diff}^k_{\mathrm{nb}}(\mathbb{R}, 0)$. □

5.3 Minimal Atlases on $\mathbb{L}$

A C^k-atlas A on $\mathbb{L}$ will be called *minimal* if it consisting of only two charts $\varphi\colon U \to \mathbb{R}$ and $\psi\colon V \to \mathbb{R}$ being homeomorphisms onto and such that $\varphi(0) = \psi(\bar{0}) = 0$.

Notice that for every such an atlas, its transition map from (U, φ) to (V, ψ)

$$\psi \circ \varphi^{-1}\colon \mathbb{R} \setminus 0 \to \mathbb{R} \setminus 0$$

is a *monotone* C^k-diffeomorphism of $\mathbb{R} \setminus 0$, i.e., it belongs to $\mathrm{Diff}^k_{\mathrm{mon}}(\mathbb{R} \setminus 0)$. The corresponding 0-extension of $\psi \circ \varphi^{-1}$ will be denoted by g_{A}, i.e.,

$$g_{\mathsf{A}} := \gamma(\psi \circ \varphi^{-1}) \in \mathcal{W}^k(\mathbb{R}, 0).$$

Recall that by definition, A is *orientable* iff its unique transition map $\psi \circ \varphi^{-1}$ preserves orientation. The latter is, evidently, equivalent to the assumption that g_{A} preserve orientations.

The following lemma is the reason for our detailed exposition of the uniqueness of C^k-structures on $\mathbb{R}$.

Lemma 16 *Every* C^k*-structure* $\mathfrak{A}$ *on* $\mathbb{L}$ *is represented by some orientable minimal atlas.* □

Proof Since U and V are open subsets homeomorphic with $\mathbb{R}$, it follows from Corollary 1 that there exist two homeomorphisms onto $\varphi\colon U \to \mathbb{R}$ and $\psi\colon V \to \mathbb{R}$ being C^k-compatible with $\mathfrak{A}$. Replacing φ and ψ with $\varphi - \varphi(0)$ and $\psi - \psi(\bar{0})$ respectively, one can achieve that $\varphi(0) = \psi(\bar{0}) = 0$. Moreover, replacing (if necessary) ψ with $-\psi$, we can assume that the transition map $\psi^{-1} \circ \varphi$ preserves orientation. As $\mathbb{L} = U \cup V$, we have that $\mathsf{A} = \{(U, \varphi), (V, \psi)\}$ is the desired orientable minimal C^k-atlas belonging to $\mathfrak{A}$. □

Theorem 4 *Let* $\mathsf{A} = \{(U, \varphi), (V, \psi)\}$ *and* $\mathsf{B} = \{(U, \bar{\varphi}), (V, \bar{\psi})\}$ *be two minimal* C^k*-atlases on* $\mathbb{L}$ *corresponding to* C^k*-structures* $\mathfrak{A}$ *and* $\mathfrak{B}$ *respectively, and* g_{A} *and* g_{B} *be* 0*-extensions of their respective transition maps* $\psi \circ \varphi^{-1}$ *and* $\bar{\psi} \circ \bar{\varphi}^{-1}$.

(1) *Then the following correspondences* $\xi^{\mathrm{fix}}_{\mathsf{A},\mathsf{B}}$ *and* $\xi^{\mathrm{ex}}_{\mathsf{A},\mathsf{B}}$ *are well-defined bijections:*

$$\xi^{\mathrm{fix}}_{\mathsf{A},\mathsf{B}} \colon \mathrm{Diff}^{\mathrm{fix}}_{\mathfrak{A},\mathfrak{B}}(\mathbb{L}) \to \mathrm{Diff}^k(\mathbb{R}, 0) \cap g_{\mathsf{B}}\, \mathrm{Diff}^k(\mathbb{R}, 0)\, g_{\mathsf{A}}^{-1}, \qquad \xi^{\mathrm{fix}}_{\mathsf{A},\mathsf{B}}(h) = \bar{\psi} \circ h \circ \psi^{-1},$$
$$\xi^{\mathrm{ex}}_{\mathsf{A},\mathsf{B}} \colon \mathrm{Diff}^{\mathrm{ex}}_{\mathfrak{A},\mathfrak{B}}(\mathbb{L}) \to \mathrm{Diff}^k(\mathbb{R}, 0) \cap g_{\mathsf{B}}\, \mathrm{Diff}^k(\mathbb{R}, 0)\, g_{\mathsf{A}}, \qquad \xi^{\mathrm{ex}}_{\mathsf{A},\mathsf{B}}(h) = \bar{\varphi} \circ h \circ \psi^{-1},$$

$$\begin{array}{ccc} V & \xrightarrow{\;h\;} & V \\ {\scriptstyle\psi}\downarrow & & \downarrow{\scriptstyle\bar{\psi}} \\ \mathbb{R} & \xrightarrow{\;\xi^{\mathrm{fix}}_{\mathsf{A},\mathsf{B}}(h)\;} & \mathbb{R} \end{array} \qquad\qquad \begin{array}{ccc} V & \xrightarrow{\;h\;} & U \\ {\scriptstyle\psi}\downarrow & & \downarrow{\scriptstyle\bar{\varphi}} \\ \mathbb{R} & \xrightarrow{\;\xi^{\mathrm{ex}}_{\mathsf{A},\mathsf{B}}(h)\;} & \mathbb{R} \end{array}$$

Thus, $\xi^{\mathrm{fix}}_{\mathsf{A},\mathsf{B}}(h)$ *is the representation of* $h \in \mathrm{Diff}^{\mathrm{fix}}_{\mathfrak{A},\mathfrak{B}}(\mathbb{L})$ *with respect to the charts* (V, ψ) *and* $(V, \bar{\psi})$*, while* $\xi^{\mathrm{ex}}_{\mathsf{A},\mathsf{B}}(h)$ *is the representation of* $h \in \mathrm{Diff}^{\mathrm{ex}}_{\mathfrak{A},\mathfrak{B}}(\mathbb{L})$ *with respect to the charts* (V, ψ) *and* $(U, \bar{\varphi})$.

(2) *If* $\bar{\psi}$ *and* ψ *mutually preserve or reverse orientations, then*

$$\xi^{\mathrm{fix}}_{\mathsf{A},\mathsf{B}}\big(\mathrm{Diff}^{+,\mathrm{fix}}_{\mathfrak{A},\mathfrak{B}}(\mathbb{L})\big) \subset \mathrm{Diff}^{k,+}(\mathbb{R}, 0), \quad \xi^{\mathrm{fix}}_{\mathsf{A},\mathsf{B}}\big(\mathrm{Diff}^{-,\mathrm{fix}}_{\mathfrak{A},\mathfrak{B}}(\mathbb{L})\big) \subset \mathrm{Diff}^{k,-}(\mathbb{R}, 0).$$

Similarly, if $\bar{\varphi}$ *and* ψ *mutually preserve or reverse orientations, then*

$$\xi^{\mathrm{ex}}_{\mathsf{A},\mathsf{B}}\big(\mathrm{Diff}^{+,\mathrm{ex}}_{\mathfrak{A},\mathfrak{B}}(\mathbb{L})\big) \subset \mathrm{Diff}^{k,+}(\mathbb{R}, 0), \quad \xi^{\mathrm{ex}}_{\mathsf{A},\mathsf{B}}\big(\mathrm{Diff}^{-,\mathrm{ex}}_{\mathfrak{A},\mathfrak{B}}(\mathbb{L})\big) \subset \mathrm{Diff}^{k,-}(\mathbb{R}, 0).$$

(3) *If* $\mathsf{A} = \mathsf{B}$*, so* $g_{\mathsf{A}} = g_{\mathsf{B}}$ *and* $\mathfrak{A} = \mathfrak{B}$*, then*

$$\xi^{\mathrm{fix}}_{\mathsf{A},\mathsf{B}} \colon \mathrm{Diff}^{\mathrm{fix}}_{\mathfrak{A}}(\mathbb{L}) \to \mathrm{Diff}^k(\mathbb{R}, 0) \cap g_{\mathsf{A}}\, \mathrm{Diff}^k(\mathbb{R}, 0)\, g_{\mathsf{A}}^{-1}$$

is an isomorphism of groups. Moreover, the group $\mathrm{Diff}^{+,\mathrm{fix}}_{\mathfrak{A}}(\mathbb{L})$ *contains a subgroup isomorphic with* $\mathrm{Diff}^k_{\mathrm{nb}}(\mathbb{R}, 0)$*, and therefore it is uncountable.*

(4) *If* $\mathrm{Diff}^{*,\bullet}_{\mathfrak{A},\mathfrak{B}}(\mathbb{L})$ *is non-empty for some* $* \in \{\pm\}$ *and* $\bullet \in \{\mathrm{fix}, \mathrm{ex}\}$*, then it is uncountable.* □

Proof (1) First we show that $\xi^{\mathrm{fix}}_{\mathsf{A},\mathsf{B}}$ is a well-defined bijection. Let $h \in \mathrm{Diff}^{\mathrm{fix}}_{\mathfrak{A},\mathfrak{B}}(\mathbb{L})$. Then $h(U) = U$ and $h(V) = V$, whence we have the following commutative diagrams:

$$\begin{array}{ccc} U & \xrightarrow{\;h\;} & U \\ {\scriptstyle\varphi}\downarrow & & \downarrow{\scriptstyle\bar{\varphi}} \\ \mathbb{R} & \xrightarrow{\;a=\bar{\varphi}\circ h\circ\varphi^{-1}\;} & \mathbb{R} \end{array} \qquad\qquad \begin{array}{ccc} V & \xrightarrow{\;h\;} & V \\ {\scriptstyle\psi}\downarrow & & \downarrow{\scriptstyle\bar{\psi}} \\ \mathbb{R} & \xrightarrow{\;b=\bar{\psi}\circ h\circ\psi^{-1}\;} & \mathbb{R} \end{array}$$

in which the lower arrows $a := \bar{\varphi} \circ h \circ \varphi^{-1}$ and $b := \bar{\psi} \circ h \circ \psi^{-1}$ are representations of h in the corresponding charts. We need to show that $\xi^{\mathrm{fix}}_{\mathsf{A},\mathsf{B}}(h) := b \in \mathrm{Diff}^k(\mathbb{R}, 0) \cap g_{\mathsf{B}}\, \mathrm{Diff}^k(\mathbb{R}, 0)\, g_{\mathsf{A}}^{-1}$ and that correspondence $h \mapsto b$ is a bijection.

The assumption that h is a C^k-diffeomorphism of $\mathbb{L}$ means that $a, b \in \mathrm{Diff}^k(\mathbb{R}, 0)$. Notice that on $\mathbb{R} \setminus 0$ we have the following identity:

$$b = \bar{\psi} \circ h \circ \psi^{-1} = (\bar{\psi} \circ \bar{\varphi}^{-1}) \circ (\bar{\varphi} \circ h \circ \varphi^{-1}) \circ (\psi \circ \varphi^{-1})^{-1}.$$

Taking 0-extensions of these diffeomorphisms we obtain that

$$b = g_{\mathsf{B}} \circ a \circ g_{\mathsf{A}}^{-1} \in g_{\mathsf{B}} \,\mathrm{Diff}^k(\mathbb{R}, 0)\, g_{\mathsf{A}}^{-1}.$$

But $b \in \mathrm{Diff}^k(\mathbb{R}, 0)$ as well, whence $b \in \mathrm{Diff}^k(\mathbb{R}, 0) \cap g_{\mathsf{B}} \,\mathrm{Diff}^k(\mathbb{R}, 0)\, g_{\mathsf{A}}^{-1}$, and thus the map $\xi^{\mathrm{fix}}_{\mathsf{A},\mathsf{B}}$ is well defined.

Moreover, since $h(0) = 0$, h is uniquely determined by its restriction to $\mathbb{L} \setminus \{0\} = V$, and therefore by the composition $b = \bar{\psi} \circ h \circ \psi^{-1}$. Hence $\xi^{\mathrm{fix}}_{\mathsf{A},\mathsf{B}}$ is injective.

Let us prove that $\xi^{\mathrm{fix}}_{\mathsf{A},\mathsf{B}}$ is surjective. Let $b \in \mathrm{Diff}^k(\mathbb{R}, 0) \cap g_{\mathsf{B}} \,\mathrm{Diff}^k(\mathbb{R}, 0)\, g_{\mathsf{A}}^{-1}$, i.e.,

$$b = g_{\mathsf{B}} \circ a \circ g_{\mathsf{A}}^{-1} = \bar{\psi} \circ \bar{\varphi}^{-1} \circ a \circ \varphi \circ \psi^{-1}$$

for some $a \in \mathrm{Diff}^k(\mathbb{R}, 0)$. The latter is equivalent to the following identity:

$$\bar{\varphi}^{-1} \circ a \circ \varphi = \bar{\psi}^{-1} \circ b \circ \psi. \tag{12}$$

Define the map

$$h\colon \mathbb{L} \to \mathbb{L}, \qquad h(x) = \begin{cases} \bar{\varphi}^{-1} \circ a \circ \varphi(x), & x \in U, \\ \bar{\psi}^{-1} \circ b \circ \psi(x), & x \in V. \end{cases} \tag{13}$$

Due to (12), the formulas for h agree on $U \cap V$, and therefore h is a continuous bijection. Moreover, we have the following commutative diagrams:

$$\begin{array}{ccc} U & \xrightarrow{\;h|_U = \bar{\varphi}^{-1} \circ a \circ \varphi\;} & U \\ {\scriptstyle\varphi}\downarrow & & \downarrow{\scriptstyle\bar{\varphi}} \\ \mathbb{R} & \xrightarrow{\;a\;} & \mathbb{R} \end{array} \qquad\qquad \begin{array}{ccc} V & \xrightarrow{\;h|_V = \bar{\psi}^{-1} \circ b \circ \psi\;} & V \\ {\scriptstyle\psi}\downarrow & & \downarrow{\scriptstyle\bar{\psi}} \\ \mathbb{R} & \xrightarrow{\;b\;} & \mathbb{R} \end{array}$$

which show that the representation of h in the charts U and V are C^k-diffeomorphisms a and b respectively. Therefore h is a C^k-diffeomorphism between $(\mathbb{L}, \mathfrak{A})$ and $(\mathbb{L}, \mathfrak{B})$. This proves that $\xi^{\mathrm{fix}}_{\mathsf{A},\mathsf{B}}$ is a bijection.

The proof for $\xi^{\mathrm{ex}}_{\mathsf{A},\mathsf{B}}$ is similar. Let us just mention that if $h \in \mathrm{Diff}^{\mathrm{ex}}_{\mathfrak{A},\mathfrak{B}}(\mathbb{L})$, then $h(U) = V$ and $h(V) = U$, whence we have the following commutative diagrams:

$$\begin{array}{ccc} U & \xrightarrow{\ h\ } & V \\ {\scriptstyle\varphi}\downarrow & & \downarrow{\scriptstyle\bar{\psi}} \\ \mathbb{R} & \xrightarrow{a=\bar{\psi}\circ h\circ\varphi^{-1}} & \mathbb{R} \end{array} \qquad\qquad \begin{array}{ccc} V & \xrightarrow{\ h\ } & U \\ {\scriptstyle\psi}\downarrow & & \downarrow{\scriptstyle\bar{\varphi}} \\ \mathbb{R} & \xrightarrow{b=\bar{\varphi}\circ h\circ\psi^{-1}} & \mathbb{R} \end{array}$$

in which the lower arrows $a := \bar{\psi} \circ h \circ \varphi^{-1}$ and $b := \bar{\varphi} \circ h \circ \psi^{-1}$ are representations of h in the corresponding charts. Then $\xi^{ex}_{\mathsf{A},\mathsf{B}}$ is defined by $\xi^{ex}_{\mathsf{A},\mathsf{B}}(h) = b$. The verification that $\xi^{ex}_{\mathsf{A},\mathsf{B}}$ is a well-defined bijection is similar to the case of $\xi^{fix}_{\mathsf{A},\mathsf{B}}$, and we leave it for the reader.

(2) This statement directly follows from the definition of bijections $\xi^{fix}_{\mathsf{A},\mathsf{B}}$ and $\xi^{ex}_{\mathsf{A},\mathsf{B}}$.

(3) Suppose $\mathsf{A} = \mathsf{B}$, so $\psi = \bar{\psi}$ and $g_{\mathsf{B}} = g_{\mathsf{A}}$. Then $g_{\mathsf{A}}\,\mathrm{Diff}^k(\mathbb{R},0)\,g_{\mathsf{A}}^{-1}$ is a conjugated to $\mathrm{Diff}^k(\mathbb{R},0)$ subgroup of $\mathcal{W}^k(\mathbb{R},0)$, whence $\mathrm{Diff}^k(\mathbb{R},0) \cap g_{\mathsf{A}}\,\mathrm{Diff}^k(\mathbb{R},0)\,g_{\mathsf{A}}^{-1}$ is also a subgroup of $\mathcal{W}^k(\mathbb{R},0)$. Moreover, $\xi^{fix}_{\mathsf{A},\mathsf{B}}(h) = g_{\mathsf{B}} \circ h \circ g_{\mathsf{B}}^{-1}$ for all $h \in \mathcal{W}^k(\mathbb{R},0)$, and therefore $\xi^{fix}_{\mathsf{A},\mathsf{B}}$ is a homomorphism of groups. Since $\xi^{fix}_{\mathsf{A},\mathsf{B}}$ is also a bijection, it is an isomorphism.

Also, since by Lemma 15, $\mathrm{Diff}^k_{nb}(\mathbb{R},0)$ is normal in $\mathcal{W}^k(\mathbb{R},0)$, we see that

$$\mathrm{Diff}^k_{nb}(\mathbb{R},0) \;=\; g_{\mathsf{A}}\,\mathrm{Diff}^k_{nb}(\mathbb{R},0)g_{\mathsf{A}}^{-1} \;\subset\; \mathrm{Diff}^k(\mathbb{R},0) \cap g_{\mathsf{A}}\,\mathrm{Diff}^k(\mathbb{R},0)\,g_{\mathsf{A}}^{-1}.$$

Hence $(\xi^{fix}_{\mathfrak{A},\mathfrak{A}})^{-1}(\mathrm{Diff}^k_{nb}(\mathbb{R},0))$ is the desired subgroup of $\mathrm{Diff}^{+,fix}_{\mathfrak{A}}(\mathbb{L})$ isomorphic to $\mathrm{Diff}^k_{nb}(\mathbb{R},0)$.

(4) It follows from Lemma 10 that for each $h \in \mathrm{Diff}^{*,\bullet}_{\mathfrak{A},\mathfrak{B}}(\mathbb{L})$ we have three bijections:

$$\begin{array}{ccccc} & & \mathrm{Diff}^{*,\bullet}_{\mathfrak{A},\mathfrak{B}}(\mathbb{L}) & & \\ & {\scriptstyle g\mapsto h\circ g}\nearrow & & \searrow{\scriptstyle k\mapsto k\circ h^{-1}} & \\ \mathrm{Diff}^{+,fix}_{\mathfrak{A}}(\mathbb{L}) & & \xrightarrow{\ g\mapsto h\circ g\circ h^{-1}\ } & & \mathrm{Diff}^{+,fix}_{\mathfrak{B}}(\mathbb{L}) \end{array}$$

Hence, if $\mathrm{Diff}^{*,\bullet}_{\mathfrak{A},\mathfrak{B}}(\mathbb{L}) \neq \varnothing$, then it is uncountable since $\mathrm{Diff}^{fix}_{\mathfrak{A}}(\mathbb{L})$ is so. □

Corollary 2 *Let $k \in \mathbb{N} \cup \{\infty\}$ and $\mathfrak{A}$ and $\mathfrak{B}$ be two C^k-structures on $\mathbb{L}$.*

(1) Then the following conditions are equivalent:

- (i^{fix}) $\mathrm{Diff}^{fix}_{\mathfrak{A},\mathfrak{B}}(\mathbb{L}) \neq \varnothing$;
- (ii^{fix}) *there exist minimal atlases* $\mathsf{A} = \{(U,\varphi),(V,\psi)\} \in \mathfrak{A}$ *and* $\mathsf{B} = \{(U,\bar{\varphi}), (V,\bar{\psi})\} \in \mathfrak{B}$ *such that* $g_{\mathsf{B}} = b \circ g_{\mathsf{A}} \circ a^{-1}$ *for some* $a, b \in \mathrm{Diff}^k(\mathbb{R},0)$, *i.e., the* 0*-extensions* g_{A} *and* g_{B} *of their respective transition maps* $\psi \circ \varphi^{-1}$ *and* $\bar{\psi} \circ \bar{\varphi}^{-1}$ *belong to the same* $\mathrm{Diff}^k(\mathbb{R},0)$*-double coset in* $\mathrm{Diff}^k(\mathbb{R},0) \backslash \mathcal{W}^k(\mathbb{R},0) \,/\, \mathrm{Diff}^k(\mathbb{R},0)$*;*
- (iii^{fix}) *property* (ii^{fix}) *holds for any pair of minimal atlases* $\mathsf{A} \in \mathfrak{A}$ *and* $\mathsf{B} \in \mathfrak{B}$.

(2) Similarly, the following conditions are also equivalent:

- (i^{ex}) $\mathrm{Diff}^{ex}_{\mathfrak{A},\mathfrak{B}}(\mathbb{L}) \neq \varnothing$;

(ii^{ex}) *there exist minimal atlases* $\mathsf{A} \in \mathfrak{A}$ *and* $\mathsf{B} \in \mathfrak{B}$ *such that* $g_{\mathsf{B}}^{-1} = b \circ g_{\mathsf{A}} \circ a^{-1}$ *for some* $a, b \in \mathrm{Diff}^k(\mathbb{R}, 0)$*;*

(iii^{ex}) *property* (ii^{ex}) *holds for any pair of minimal atlases* $\mathsf{A} \in \mathfrak{A}$ *and* $\mathsf{B} \in \mathfrak{B}$.

(3) Hence, the following conditions are equivalent as well:

(i) $\mathrm{Diff}_{\mathfrak{A},\mathfrak{B}}(\mathbb{L}) \equiv \mathrm{Diff}^{fix}_{\mathfrak{A},\mathfrak{B}}(\mathbb{L}) \sqcup \mathrm{Diff}^{ex}_{\mathfrak{A},\mathfrak{B}}(\mathbb{L}) \neq \varnothing$*;*

(ii) *there exist minimal atlases* $\mathsf{A} \in \mathfrak{A}$ *and* $\mathsf{B} \in \mathfrak{B}$ *such that either* $g_{\mathsf{B}} = b \circ g_{\mathsf{A}} \circ a^{-1}$ *or* $g_{\mathsf{B}}^{-1} = b \circ g_{\mathsf{A}} \circ a^{-1}$ *for some* $a, b \in \mathrm{Diff}^k(\mathbb{R}, 0)$*, i.e.,* g_{A} *and* g_{B} *belong to the same* $(\mathrm{Diff}^k(\mathbb{R}, 0), \pm)$*-double coset in* $\mathrm{Diff}^k(\mathbb{R}, 0) \setminus \mathcal{W}^k(\mathbb{R}, 0)^{\pm 1} / \mathrm{Diff}^k(\mathbb{R}, 0)$*;*

(iii) *property (ii) holds for any pair of minimal atlases* $\mathsf{A} \in \mathfrak{A}$ *and* $\mathsf{B} \in \mathfrak{B}$. □

Proof (1) The implication (iii^{fix}) $\Rightarrow$ (ii^{fix}) is trivial.

(i^{fix}) $\Rightarrow$ (iii^{fix}). Suppose $\mathrm{Diff}^{fix}_{\mathfrak{A},\mathfrak{B}}(\mathbb{L}) \neq \varnothing$ and let $\mathsf{A} \in \mathfrak{A}$ and $\mathsf{B} \in \mathfrak{B}$ be any minimal atlases. Then by Theorem 4, we have a bijection

$$\xi^{fix}_{\mathsf{A},\mathsf{B}} \colon \mathrm{Diff}^{fix}_{\mathfrak{A},\mathfrak{B}}(\mathbb{L}) \equiv \mathrm{Diff}^k(\mathbb{R}, 0) \cap g_{\mathsf{B}}\, \mathrm{Diff}^k(\mathbb{R}, 0)\, g_{\mathsf{A}}^{-1}.$$

Hence, the latter intersection is non-empty, i.e., $b = g_{\mathsf{B}} \circ a \circ g_{\mathsf{A}}^{-1}$ for some $a, b \in \mathrm{Diff}^k(\mathbb{R}, 0)$, which is the same as $g_{\mathsf{B}} = b \circ g_{\mathsf{A}} \circ a^{-1}$.

(ii^{fix}) $\Rightarrow$ (i^{fix}) Suppose there exist some minimal atlases $\mathsf{A} \in \mathfrak{A}$ and $\mathsf{B} \in \mathfrak{B}$ such that $g_{\mathsf{B}} = b \circ g_{\mathsf{A}} \circ a^{-1}$ for some $a, b \in \mathrm{Diff}^k(\mathbb{R}, 0)$. Then

$$b = g_{\mathsf{B}} \circ a \circ g_{\mathsf{A}}^{-1} \in \mathrm{Diff}^k(\mathbb{R}, 0) \cap g_{\mathsf{B}}\, \mathrm{Diff}^k(\mathbb{R}, 0)\, g_{\mathsf{A}}^{-1} \neq \varnothing.$$

Hence, by Theorem 4, $\mathrm{Diff}^{fix}_{\mathfrak{A},\mathfrak{B}}(\mathbb{L}) \neq \varnothing$ as well.

The proof of (2) is similar, and (3) is just a combination of (1) and (2). □

5.4 *Classification of C^k-Structures on* $\mathbb{L}$

Let $C^k(\mathbb{L})$ be the set of all C^k-structures on $\mathbb{L}$. Then the group $\mathcal{H}(\mathbb{L})$ of homeomorphisms of $\mathbb{L}$ acts on $C^k(\mathbb{L})$, as described in Lemma 8(3b). For every $\mathfrak{A} \in C^k(\mathbb{L})$ let

- $[\mathfrak{A}]$ be the orbit of $\mathfrak{A}$ with respect to that action of $\mathcal{H}(\mathbb{L})$;
- $[\mathfrak{A}]^{fix}$ be the orbit of $\mathfrak{A}$ with respect to the induced action of the subgroup $\mathcal{H}^{fix}(\mathbb{L})$ consisting of homeomorphisms fixing 0 and $\bar{0}$.

Thus $[\mathfrak{A}]$ consist of all C^k-structures on $\mathbb{L}$ being C^k-diffeomorphic to $\mathfrak{A}$, while $[\mathfrak{A}]^{fix}$ is a subset of $[\mathfrak{A}]$ consisting of C^k-structures being C^k-diffeomorphic to $\mathfrak{A}$ via a diffeomorphism fixing 0 and $\bar{0}$.

Denote by $C^k(\mathbb{L})/\mathcal{H}(\mathbb{L})$ and $C^k(\mathbb{L})/\mathcal{H}^{fix}(\mathbb{L})$ the corresponding orbit spaces. As a consequence we get the following theorem clarifying the bijections mentioned in Theorem 1.

Theorem 5 *There are well defined* bijections

- $\sigma\colon C^k(\mathbb{L})/\mathcal{H}(\mathbb{L}) \longrightarrow \mathrm{Diff}^k(\mathbb{R},0) \setminus \mathcal{W}^k(\mathbb{R},0)^{\pm 1} \,/\, \mathrm{Diff}^k(\mathbb{R},0)$,

$$\sigma([\mathfrak{A}]) = \mathrm{Diff}^k(\mathbb{R},0)\, g_{\mathsf{A}}^{\pm 1}\, \mathrm{Diff}^k(\mathbb{R},0), \text{ where } \mathsf{A} \in \mathfrak{A} \text{ is any minimal atlas,}$$

- $\sigma^{\mathrm{fix}}\colon C^k(\mathbb{L})/\mathcal{H}^{\mathrm{fix}}(\mathbb{L}) \longrightarrow \mathrm{Diff}^k(\mathbb{R},0) \setminus \mathcal{W}^k(\mathbb{R},0) \,/\, \mathrm{Diff}^k(\mathbb{R},0)$,

$$\sigma^{\mathrm{fix}}([\mathfrak{A}]^{\mathrm{fix}}) = \mathrm{Diff}^k(\mathbb{R},0)\, g_{\mathsf{A}}\, \mathrm{Diff}^k(\mathbb{R},0), \text{ where } \mathsf{A} \in \mathfrak{A} \text{ is any minimal atlas,}$$

associating to the isomorphism classes $[\mathfrak{A}]$ *and* $[\mathfrak{A}]^{\mathrm{fix}}$ *of a* C^k*-structure* $\mathfrak{A}$ *on* $\mathbb{L}$ *respectively* $\mathrm{Diff}^k(\mathbb{R},0)$*- and* $(\mathrm{Diff}^k(\mathbb{R},0),\pm)$*-double cosets in* $\mathcal{W}^k(\mathbb{R},0)$ *of the* 0*-extension* g_{A} *of the transition map of* any *minimal atlas* $\mathsf{A} \in \mathfrak{A}$. □

Proof It is just a rephrasing of Corollary 2. Namely, the statement that σ is a bijection means that two C^k-structures $\mathfrak{A}$ and $\mathfrak{B}$ on $\mathbb{L}$ are C^k-diffeomorphic iff for any pair of minimal atlases $\mathsf{A} \in \mathfrak{A}$ and $\mathsf{B} \in \mathfrak{B}$ the 0-extensions g_{A} and g_{B} of their respective transition maps belong to the same $(\mathrm{Diff}^k(\mathbb{R},0),\pm)$-double coset in $\mathcal{W}^k(\mathbb{R},0)$. This is exactly the equivalence of (i) and (iii) of Corollary 2.

Similarly, existence of the bijection σ^{fix} is the same as the equivalence $(\mathrm{i}^{\mathrm{fix}}) \Leftrightarrow (\mathrm{iii}^{\mathrm{fix}})$. □

5.5 Special Minimal Atlases

Our next aim is to provide examples of C^k-structures $\mathfrak{A}$ and $\mathfrak{B}$ on $\mathbb{L}$ for which either of sets $\mathrm{Diff}^{\mathrm{fix}}_{\mathfrak{A},\mathfrak{B}}(\mathbb{L})$ and $\mathrm{Diff}^{\mathrm{ex}}_{\mathfrak{A},\mathfrak{B}}(\mathbb{L})$ can be empty or non-empty. For that reason it will be convenient to work only with C^k-structures admitting *special* minimal atlases defined as follows.

Recall that $U = \mathbb{R}$ and $V = (\mathbb{R} \setminus \{0\}) \cup \{\bar{0}\}$ by the construction of $\mathbb{L}$. Hence, we have the following homeomorphisms: the identity map $\mathrm{id}_u\colon U \to \mathbb{R}$, and

$$\mathrm{id}_v\colon V \to \mathbb{R}, \qquad \mathrm{id}_v(x) = \begin{cases} 0, & x = \bar{0}, \\ x = \mathrm{id}_u(x), & x \in V \setminus \bar{0}. \end{cases}$$

In particular, $\mathrm{id}_u = \mathrm{id}_v$ on $U \setminus 0 = V \setminus \bar{0}$.

Then for each $h \in \mathcal{W}^k(\mathbb{R},0)$ one can define the following minimal atlas

$$\mathsf{S}(h) := \{(U, \mathrm{id}_u), (V, h \circ \mathrm{id}_v)\}$$

whose transition map is

$$h \circ \mathrm{id}_v \circ \mathrm{id}_u^{-1}|_{\mathbb{R}\setminus 0} = h|_{\mathbb{R}\setminus 0}.$$

Hence, its 0-extension is h, i.e., $g_{\mathsf{S}(h)} = h$.

The atlas $\mathsf{S}(h)$ will be called the *special minimal atlas of* $h \in \mathcal{W}^k(\mathbb{R}, 0)$. Denote by $\mathfrak{S}(h)$ the corresponding C^k-structure on $\mathbb{L}$ defined by the atlas $\mathsf{S}(h)$, so it is determined by h. The following lemma collects simple facts about special atlases.

Lemma 17 (1) *Let* $\mathsf{A} = \{(U, \varphi), (V, \psi)\}$ *be a minimal atlas on* $\mathbb{L}$ *such that* $\varphi = \mathrm{id}_u$. *Then* A *is the special minimal atlas corresponding to the homeomorphism* $\psi \circ \mathrm{id}_v^{-1} \in \mathcal{W}^k(\mathbb{R}, 0)$, *i.e.,* $\mathsf{A} = \mathsf{S}(\psi \circ \mathrm{id}_v^{-1})$.

(2) *For a* C^k*-structure* $\mathfrak{A}$ *on* $\mathbb{L}$ *the following conditions are equivalent:*

   (a) $\mathfrak{A}$ *has a special minimal* C^k*-atlas;*
   (b) *there exists a minimal atlas* $\mathsf{A} = \{(U, \varphi), (V, \psi)\} \in \mathfrak{A}$ *such that* $\varphi\colon U = \mathbb{R} \to \mathbb{R}$ *is a* C^k*-diffeomorphism;*
   (c) *for every minimal atlas* $\mathsf{A} = \{(U, \varphi), (V, \psi)\} \in \mathfrak{A}$ *we have that* $\varphi\colon U = \mathbb{R} \to \mathbb{R}$ *is a* C^k*-diffeomorphism.*

(3) *Let* $\mathfrak{A}$ *be a* C^k*-structure on* $\mathbb{L}$, $\mathsf{A} = \{(U, \varphi), (V, \psi)\} \in \mathfrak{A}$ *some minimal atlas, and* g_{A} *be the* 0*-extension of its transition map* $\psi \circ \varphi^{-1}$. *Then the* C^k*-structures* $\mathfrak{A}$ *and* $\mathfrak{S}(g_{\mathsf{A}})$ *are diffeomorphic via a preserving orientation diffeomorphism* h *fixing* 0 *and* $\bar{0}$, *i.e.,* $\mathrm{Diff}^{+,\mathrm{fix}}_{\mathfrak{A},\mathfrak{S}(g_{\mathsf{A}})}(\mathbb{L}) \neq \varnothing$. *For instance, if* φ *preserves orientation, then the following map* $h\colon \mathbb{L} \to \mathbb{L}$ *defined by*

$$h|_U = \mathrm{id}_u^{-1} \circ \varphi\colon U \to U \qquad \textit{and} \qquad h(\bar{0}) = \bar{0}$$

belongs to $\mathrm{Diff}^{+,\mathrm{fix}}_{\mathfrak{A},\mathfrak{S}(g_{\mathsf{A}})}(\mathbb{L})$.

(4) *Let* $g, h \in \mathrm{Diff}^k(\mathbb{R}, 0)$. *Then* $\mathfrak{S}(h) = \mathfrak{S}(g)$ *if and only if* $g \circ h^{-1} \in \mathrm{Diff}^k(\mathbb{R}, 0)$. *In particular, there is a bijection between* C^k*-structures on* $\mathbb{L}$ *admitting special minimal atlas and* right *cosets* $\mathrm{Diff}^k(\mathbb{R}, 0) \setminus \mathcal{W}^k(\mathbb{R}, 0)$.

(5) *For each* $g, h \in \mathcal{W}^k(\mathbb{R}, 0)$ *there are bijections:*

$$\mathrm{Diff}^{\mathrm{fix}}_{\mathfrak{S}(g),\mathfrak{S}(h)}(\mathbb{L}) \xrightarrow{\xi_{\mathsf{S}(g),\mathsf{S}(h)}} \mathrm{Diff}^k(\mathbb{R}, 0) \cap h\, \mathrm{Diff}^k(\mathbb{R}, 0)\, g^{-1} \xleftarrow{\xi_{\mathsf{S}(g^{-1}),\mathsf{S}(h)}} \mathrm{Diff}^{\mathrm{ex}}_{\mathfrak{S}(g^{-1}),\mathfrak{S}(h)}(\mathbb{L}).$$

(6) *If* $b \in \mathcal{W}^k(\mathbb{R}, 0)$ *but* $b \notin \mathrm{Diff}^k(\mathbb{R}, 0)$, *then the* C^k*-structures* $\mathfrak{S}(\mathrm{id}_{\mathbb{R}})$ *and* $\mathfrak{S}(b)$ *are not* C^k*-diffeomorphic, i.e.,* $\mathrm{Diff}_{\mathfrak{S}(\mathrm{id}_{\mathbb{R}}),\mathfrak{S}(b)}(\mathbb{L}) = \varnothing$. □

Proof Statements (1) is trivial, while (5) follows from Theorem 4.

(2) The implication (c)⇒(b) is also trivial.

(b)⇒(a) Let $\mathsf{A} = \{(U, \varphi), (V, \psi)\} \in \mathfrak{A}$ be a C^k-atlas such that $\varphi\colon U = \mathbb{R} \to \mathbb{R}$ is a C^k-diffeomorphism. Then $\{(U, \varphi^{-1} \circ \varphi = \mathrm{id}_u), (V, \psi)\}$ is a special minimal C^k-atlas C^k-compatible with A, and therefore belonging to $\mathfrak{B}$.

(a)⇒(c) Let $\mathsf{S}(a) = \{(U, \mathrm{id}_u), (V, a \circ \mathrm{id}_v)\} \in \mathfrak{A}$ be a special minimal atlas, and $\mathsf{A} = \{(U, \varphi), (V, \psi)\}$ be any other minimal atlas in $\mathfrak{A}$. Then they must be C^k-compatible, i.e., $\varphi \circ \mathrm{id}_u^{-1} = \varphi$ is a C^k-diffeomorphism.

(3) By Theorem 4 we have a bijection $\mathrm{Diff}^{\mathrm{fix}}_{\mathfrak{A},\mathfrak{S}(g_{\mathsf{A}})}(\mathbb{L}) \equiv \mathrm{Diff}^k(\mathbb{R},0) \cap g_{\mathsf{A}}\,\mathrm{Diff}^k(\mathbb{R},0) g_{\mathsf{A}}^{-1}$. The latter intersection contains $\mathrm{id}_{\mathbb{R}}$, and therefore it is non-empty, whence $\mathfrak{A}$ and $\mathfrak{S}(g_{\mathsf{A}})$ are C^k-diffeomorphic via a diffeomorphism fixing 0 and $\bar{0}$.

(4) Consider the corresponding special minimal atlases

$$\mathsf{S}(g) := \{(U, \mathrm{id}_u), (V, g \circ \mathrm{id}_v)\}, \qquad \mathsf{S}(h) := \{(U, \mathrm{id}_u), (V, h \circ \mathrm{id}_v)\}.$$

Since they have the same first chart (U, id_u), the C^k-structures $\mathsf{S}(g)$ and $\mathsf{S}(h)$ coincide if and only if their second charts are C^k-compatible. The latter means that the corresponding transition map $(h \circ \mathrm{id}_v) \circ (g \circ \mathrm{id}_v)^{-1} = h \circ g^{-1} \in \mathrm{Diff}^k(\mathbb{R},0)$.

(6) By Theorem 4 we have the following bijections:

$$\begin{aligned}\mathrm{Diff}^{\mathrm{fix}}_{\mathfrak{S}(\mathrm{id}_{\mathbb{R}}),\mathfrak{S}(b)}(\mathbb{L}) &\overset{\xi_{\mathsf{S}(\mathrm{id}_{\mathbb{R}}),\mathsf{S}(b)}}{\equiv} \mathrm{Diff}^k(\mathbb{R},0) \cap b\,\mathrm{Diff}^k(\mathbb{R},0)\,\mathrm{id}_{\mathbb{R}}^{-1} = \mathrm{Diff}^k(\mathbb{R},0) \cap b\,\mathrm{Diff}^k(\mathbb{R},0) = \\ &= \mathrm{Diff}^k(\mathbb{R},0) \cap b\,\mathrm{Diff}^k(\mathbb{R},0)\,\mathrm{id}_{\mathbb{R}} \overset{(\xi_{\mathsf{S}(\mathrm{id}_{\mathbb{R}}^{-1}),\mathsf{S}(b)})^{-1}}{\equiv} \mathrm{Diff}^{\mathrm{ex}}_{\mathfrak{S}(\mathrm{id}_{\mathbb{R}}),\mathfrak{S}(b)}(\mathbb{L}).\end{aligned} \tag{14}$$

Since $b \notin \mathrm{Diff}^k(\mathbb{R},0)$, the adjacent classes $\mathrm{Diff}^k(\mathbb{R},0)$ and $b\,\mathrm{Diff}^k(\mathbb{R},0)$ do not intersect each other, so all the sets in (14) are empty. Hence,

$$\mathrm{Diff}_{\mathfrak{S}(\mathrm{id}_{\mathbb{R}}),\mathfrak{S}(b)}(\mathbb{L}) = \mathrm{Diff}^{\mathrm{fix}}_{\mathfrak{S}(\mathrm{id}_{\mathbb{R}}),\mathfrak{S}(b)}(\mathbb{L}) \sqcup \mathrm{Diff}^{\mathrm{ex}}_{\mathfrak{S}(\mathrm{id}_{\mathbb{R}}),\mathfrak{S}(b)}(\mathbb{L}) = \varnothing.$$

Therefore, there are no diffeomorphisms between $\mathfrak{S}(\mathrm{id}_{\mathbb{R}})$ and $\mathfrak{S}(b)$ (neiter fixing 0 and $\bar{0}$ nor exchanging them). □

Statement (3) of Lemma 17 shows that in order to classify C^k-structures on $\mathbb{L}$ up to a C^k-diffeomorphism, it suffices to consider only C^k-structures having special minimal atlases. Also, due to statement (6), $\mathbb{L}$ admits at least two non-diffeomorphic C^k-structures.

Our final step is to show that in fact, there are uncountably many pair-wise non-diffeomorphic C^k-structures on $\mathbb{L}$.

Recall that for each $a > 0$ we have defined a homeomorphism w_a by formula (2), so $w_a(x) = x$ for $x \le 0$, and $w_a(x) = ax$ for $x \ge 0$. The following simple statement allows to classify C^k-structures $\mathfrak{S}(w_a)$ on $\mathbb{L}$ corresponding to special minimal atlases $\mathsf{S}(w_a)$ of w_a for all $a > 0$.

Lemma 18 *Let $a, b > 0$, $f \in \mathrm{Diff}^k(\mathbb{R},0)$, $\tau_1 = f'(0)$, and*

$$q := w_b \circ f \circ w_a \in \mathcal{W}^k(\mathbb{R},0).$$

(a) *Suppose $\tau_1 > 0$. Then q is C^1 if and only if $a = 1/b$.*
(b) *Let $\tau_1 < 0$. Then q is C^1 if and only if $a = b$.*
(c) *Suppose q is C^1 and let $n \in \mathbb{N}$ be such that $2 \le n \le k$. Assume also that $a, b \ne 1$. Then q is C^n if and only if $f''(0) = \cdots = f^{(n)}(0) = 0$.* □

Proof Since f is of class C^k, we have that $f(x) = \tau_1 x + \tau_2 x^2 + \cdots + \tau_n x^n + g(x)$, where $\tau_i = \frac{1}{i!} f^{(i)}(0)$, $i = 1, \ldots, n$, and $g\colon \mathbb{R} \to \mathbb{R}$ is a C^k-function such that $\lim\limits_{x\to 0}\big(g(x)/x^n\big) = 0$. It follows that if $\tau_1 > 0$, then

$$q(x) = \begin{cases} f(x) = \tau_1 x + \tau_2 x^2 + \cdots \tau_n x^n + g(x), & x \le 0, \\ bf(ax) = ba\tau_1 x + ba^2\tau_2 x^2 + \cdots ba^n \tau_n x^n + bg(ax), & x \ge 0. \end{cases} \tag{15}$$

On the other hand, if $\tau_1 < 0$, then

$$q(x) = \begin{cases} bf(x) = b\tau_1 x + b\tau_2 x^2 + \cdots b\tau_n x^n + bg(x), & x \le 0, \\ f(ax) = a\tau_1 x + a^2\tau_2 x^2 + \cdots a^n \tau_n x^n + g(ax), & x \ge 0. \end{cases} \tag{16}$$

Thus if $\tau_1 > 0$, then due to (15), q is C^1 iff $ba = 1$. Moreover, in that case q is C^n iff $a^{i-1}\tau_i = \tau_i$ for $i = 2, \ldots, n$. Since $a \ne 0$, the latter is possible iff $\tau_i = 0$ for $i = 2 \ldots, n$. This proves (a) and a part of (c) for $\tau_1 > 0$.

Similarly, if $\tau_1 < 0$, then due to (16), q is C^1 iff $b = a$. Moreover, again that case q is C^n iff $a^{i-1}\tau_i = \tau_i$ for $i = 2, \ldots, n$, which is again equivalent to the assumption that $\tau_i = 0$ for $i = 2 \ldots, n$. This proves (b) and completes (c) for $\tau_1 < 0$. □

Corollary 3 *For all $a, b \ne 1$ we have that*

$$\mathrm{Diff}^k(\mathbb{R}, 0) \cap w_b\, \mathrm{Diff}^k(\mathbb{R}, 0) w_a = \begin{cases} \varnothing, & a \ne b, 1/b, \\ \mathcal{E}^{k,+}(\mathbb{R}, 0), & ab = 1, \\ \mathcal{E}^{k,-}(\mathbb{R}, 0), & a = b. \end{cases}$$

Proof (1) If $a \ne b, 1/b$, then by Lemma 18 for any $f \in \mathrm{Diff}^k(\mathbb{R}, 0)$ the homeomorphism $q = w_b \circ f \circ w_a$ is not even C^1. In particular, $\mathrm{Diff}^k(\mathbb{R}, 0) \cap w_b\, \mathrm{Diff}^k(\mathbb{R}, 0) w_a = \varnothing$.

(2) Suppose $a = 1/b$. Again, let $f \in \mathrm{Diff}^k(\mathbb{R}, 0)$ and $q = w_b \circ f \circ w_a$.

If $f \in \mathrm{Diff}^{k,-}(\mathbb{R}, 0)$, then by (b), q is not C^1 since $a \ne b$. On the other hand, if $f \in \mathrm{Diff}^{k,+}(\mathbb{R}, 0)$, then by (a) and (c), q is C^k iff $f \in \mathcal{E}^{k,+}(\mathbb{R}, 0)$. In that case, it also follows from (15) that $q \in \mathcal{E}^{k,+}(\mathbb{R}, 0)$. In other words,

$$\mathrm{Diff}^k(\mathbb{R}, 0) \cap w_b\, \mathrm{Diff}^k(\mathbb{R}, 0) w_a \;=\; \mathrm{Diff}^k(\mathbb{R}, 0) \cap w_b \mathcal{E}^{k,+}(\mathbb{R}, 0) w_a \;\subset\; \mathcal{E}^{k,+}(\mathbb{R}, 0).$$

To prove the inverse inclusion take any $\bar{q} \in \mathcal{E}^{k,+}(\mathbb{R}, 0)$, and let $\bar{f} = w_{1/b} \circ \bar{q} \circ w_{1/a}$. Since $1/a = 1/(1/b)$, the already proved part of case 2) applied to $\bar{q}$ implies that $\bar{f} \in \mathcal{E}^{k,+}(\mathbb{R}, 0)$. Hence,

$$\bar{q} = w_b \circ \bar{f} \circ w_a \in \mathrm{Diff}^k(\mathbb{R}, 0) \cap w_b \mathcal{E}^{k,+}(\mathbb{R}, 0) w_a.$$

(3) The proof that $\mathrm{Diff}^k(\mathbb{R}, 0) \cap w_b\, \mathrm{Diff}^k(\mathbb{R}, 0) w_a = \mathcal{E}^{k,-}(\mathbb{R}, 0)$ for $a = b \ne 1$ is similar. □

6 Proof of Theorem 2

Denote by $\mathfrak{S}(a)$ the C^k-structure on $\mathbb{L}$ corresponding to the special minimal atlas $\mathsf{S}(w_a)$ of the homeomorphism w_a. We need to describe the sets of diffeomorphisms between C^k-structures $\mathfrak{S}(a)$ and $\mathfrak{S}(b)$ for all $a, b > 0$. Let

$$\mathsf{S}(w_a) = \{(U, \mathrm{id}_u), (V, w_a \circ \mathrm{id}_v)\}, \qquad \mathsf{S}(w_b) = \{(U, \mathrm{id}_u), (V, w_b \circ \mathrm{id}_v)\}$$

be special minimal atlases of w_a and w_b respectively. Then by Theorem 4 we have the following bijections:

$$\xi^{\mathrm{fix}}\colon \mathrm{Diff}^{\mathrm{fix}}_{\mathfrak{S}(a),\mathfrak{S}(b)}(\mathbb{L}) \to \mathrm{Diff}^k(\mathbb{R},0) \cap w_b \,\mathrm{Diff}^k(\mathbb{R},0) w_a^{-1}, \quad \xi^{\mathrm{fix}}(h) = w_b \circ \mathrm{id}_v \circ h \circ \mathrm{id}_v^{-1} \circ w_a^{-1},$$
$$\xi^{\mathrm{ex}}\colon \mathrm{Diff}^{\mathrm{fix}}_{\mathfrak{S}(a),\mathfrak{S}(b)}(\mathbb{L}) \to \mathrm{Diff}^k(\mathbb{R},0) \cap w_b \,\mathrm{Diff}^k(\mathbb{R},0) w_a, \quad \xi^{\mathrm{ex}}(h) = \mathrm{id}_u \circ h \circ \mathrm{id}_v^{-1} \circ w_a^{-1}$$

given by the lower arrows of the corresponding diagrams:

$$\begin{array}{ccc} V & \xrightarrow{\;h\;} & V \\ {\scriptstyle w_a\circ \mathrm{id}_v}\downarrow & & \downarrow{\scriptstyle w_b\circ \mathrm{id}_v} \\ \mathbb{R} & \xrightarrow{\;\xi^{\mathrm{fix}}(h)\;} & \mathbb{R} \end{array} \qquad\qquad \begin{array}{ccc} V & \xrightarrow{\;h\;} & U \\ {\scriptstyle w_a\circ \mathrm{id}_v}\downarrow & & \downarrow{\scriptstyle \mathrm{id}_u} \\ \mathbb{R} & \xrightarrow{\;\xi^{\mathrm{ex}}(h)\;} & \mathbb{R} \end{array}$$

and thus associating to each $h \in \mathrm{Diff}_{\mathfrak{S}(a),\mathfrak{S}(b)}(\mathbb{L})$ its representation from chart V. Moreover, Corollary 3 gives an explicit description of the images of ξ^{fix} and ξ^{ex}.

(4) If $a \neq b, 1/b$, then by Corollary 3 the intersections $\mathrm{Diff}^k(\mathbb{R},0) \cap w_b \,\mathrm{Diff}^k(\mathbb{R},0) w_a^{\pm 1}$ are empty, whence $\mathrm{Diff}_{\mathfrak{S}(a),\mathfrak{S}(b)}(\mathbb{L}) = \varnothing$ as well.

(1) Let $a = 1$, so $w_a = \mathrm{id}_{\mathbb{R}}$. Then

$$w_1 \,\mathrm{Diff}^k(\mathbb{R},0) w_1^{-1} = \mathrm{Diff}^k(\mathbb{R},0) \cap w_1 \,\mathrm{Diff}^k(\mathbb{R},0) w_1 = \mathrm{Diff}^k(\mathbb{R},0),$$

whence we above bijections reduce to the following ones:

$$\xi^{\mathrm{fix}}\colon \mathrm{Diff}^{\mathrm{fix}}_{\mathfrak{S}(1)}(\mathbb{L}) \to \mathrm{Diff}^k(\mathbb{R},0), \qquad \xi^{\mathrm{ex}}\colon \mathrm{Diff}^{\mathrm{ex}}_{\mathfrak{S}(1)}(\mathbb{L}) \to \mathrm{Diff}^k(\mathbb{R},0).$$

Let $\zeta : \mathbb{L} \to \mathbb{L}$ be the involution exchanging 0 and $\bar{0}$ and fixed on $\mathbb{R} \setminus 0$. One easily checks ζ is a self-diffeomorphism of $\mathrm{Diff}^{\mathrm{ex}}_{\mathfrak{S}(1)}(\mathbb{L})$. Namely, the representations of ζ with respect to the special minimal atlas $\mathsf{S}(w_1) = \{(U, \mathrm{id}_u), (V, w_1 \circ \mathrm{id}_v = \mathrm{id}_v)\}$ are given by the following commutative diagrams in which both lower arrows are $\mathrm{id}_{\mathbb{R}}$:

$$\begin{array}{ccc} U & \xrightarrow{\;\zeta\;} & V \\ {\scriptstyle \mathrm{id}_u}\downarrow & & \downarrow{\scriptstyle \mathrm{id}_v} \\ \mathbb{R} & \xrightarrow{\;\mathrm{id}_{\mathbb{R}}\;} & \mathbb{R} \end{array} \qquad\qquad \begin{array}{ccc} V & \xrightarrow{\;\zeta\;} & U \\ {\scriptstyle \mathrm{id}_v}\downarrow & & \downarrow{\scriptstyle \mathrm{id}_u} \\ \mathbb{R} & \xrightarrow{\;\mathrm{id}_{\mathbb{R}}\;} & \mathbb{R} \end{array}$$

In particular, $\xi^{\mathrm{ex}}(\zeta) = \mathrm{id}_{\mathbb{R}}$.

Moreover, as noted in Lemma 13, ζ commutes with all homeomorphisms of $\mathbb{L}$, and in particular with $\mathrm{Diff}^{\mathrm{fix}}_{\mathfrak{S}(1)}(\mathbb{L})$. Hence the map $\eta\colon \mathrm{Diff}^k(\mathbb{R}, 0) \times \mathbb{Z}_2 \to \mathrm{Diff}(\mathbb{L})$, $\eta(h, \delta) \mapsto \zeta^{\delta} \circ h$, is an isomorphism of groups. It is also evident, that is the identities (3) hold.

(2) Suppose $a \neq 1$. Note also that $w_a^{-1} = w_{1/a}$. Then by Theorem 4 and Corollary 3 the above bijections reduce to the following ones

$$\begin{aligned}
\xi^{\mathrm{fix}}\colon \mathrm{Diff}^{\mathrm{fix}}_{\mathfrak{S}(a)}(\mathbb{L}) &\to \mathrm{Diff}^k(\mathbb{R}, 0) \cap w_a \,\mathrm{Diff}^k(\mathbb{R}, 0) w_{1/a} = \mathcal{E}^{k,+}(\mathbb{R}, 0),\\
\xi^{\mathrm{ex}}\colon \mathrm{Diff}^{\mathrm{ex}}_{\mathfrak{S}(a)}(\mathbb{L}) &\to \mathrm{Diff}^k(\mathbb{R}, 0) \cap w_a \,\mathrm{Diff}^k(\mathbb{R}, 0) w_a = \mathcal{E}^{k,-}(\mathbb{R}, 0),
\end{aligned}$$

which send preserving (resp. reversing) orientation diffeomorphisms of $\mathbb{L}$ to diffeomorphisms of $\mathbb{R}$ of the same type. Hence $\mathrm{Diff}^{-,\mathrm{fix}}_{\mathfrak{S}(a)}(\mathbb{L}) = \mathrm{Diff}^{+,\mathrm{ex}}_{\mathfrak{S}(a)}(\mathbb{L}) = \varnothing$, and

$$\mathrm{Diff}_{\mathfrak{S}(a)}(\mathbb{L}) = \mathrm{Diff}^{+,\mathrm{fix}}_{\mathfrak{S}(a)}(\mathbb{L}) \sqcup \mathrm{Diff}^{-,\mathrm{ex}}_{\mathfrak{S}(a)}(\mathbb{L}).$$

Finally, consider the following homeomorphism $p\colon \mathbb{L} \to \mathbb{L}$:

$$p(x) = \begin{cases} -x/\sqrt{a}, & x \in (-\infty; 0),\\ \bar{0}, & x = 0,\\ 0, & x = \bar{0},\\ -x\sqrt{a}, & x \in (0; \infty). \end{cases}$$

It exchanges 0 and $\bar{0}$ and one easily checks that p has order 2. We claim that p is the desired C^k-diffeomorphism of $\mathfrak{S}(a)$ belonging to $\mathrm{Diff}^{-,\mathrm{ex}}_{\mathfrak{S}(a)}(\mathbb{L})$.

To check that, let us will compute representations of p in each of the charts of the special minimal atlas $\mathsf{S}(w_a) = \{(U, \mathrm{id}_u), (V, w_a \circ \mathrm{id}_v)\}$ of w_a. Note that p exchanges the connected components of $\mathbb{L} \setminus \{0, \bar{0}\} = \mathbb{R} \setminus 0$. Then the representation of p with respect to the charts (U, id_u) and $(V, w_a \circ \mathrm{id}_v)$ on the components $(-\infty; 0)$ and $(0; \infty)$ are given by the following two diagrams:

$$\begin{array}{ccc}
(-\infty; 0) & \xrightarrow{p(x)=-x/\sqrt{a}} & (0; +\infty)\\
{\scriptstyle \mathrm{id}_u}\downarrow & & \downarrow{\scriptstyle w_a \circ \mathrm{id}_v(x)=ax}\\
(-\infty; 0) & \xrightarrow{x \mapsto -\sqrt{a}x} & (0; +\infty)
\end{array}
\qquad
\begin{array}{ccc}
(0; \infty) & \xrightarrow{p(x)=-\sqrt{a}x} & (-\infty; 0)\\
{\scriptstyle \mathrm{id}_u}\downarrow & & \downarrow{\scriptstyle w_a \circ \mathrm{id}_v(x)=x}\\
(0; \infty) & \xrightarrow{x \mapsto -\sqrt{a}x} & (-\infty; 0).
\end{array}$$

The formulas for the lower arrows are the same, and thus the representation of p with respect to the charts (U, id_u) and $(V, w_a \circ \mathrm{id}_v)$ is the map $x \mapsto -\sqrt{a}x$.

Similarly, the representation of p with respect to the charts $(V, w_a \circ \mathrm{id}_v)$ and (U, id_u) on the components $(-\infty; 0)$ and $(0; \infty)$ are given by

$$\begin{array}{ccc} (-\infty;0) & \xrightarrow{p(x)=-x/\sqrt{a}} & (0;+\infty) \\ {\scriptstyle w_a\circ \mathrm{id}_v(x)=x}\downarrow & & \downarrow{\scriptstyle \mathrm{id}_u} \\ (-\infty;0) & \xrightarrow{x\mapsto -x/\sqrt{a}} & (0;+\infty) \end{array} \qquad \begin{array}{ccc} (0;\infty) & \xrightarrow{p(x)=-\sqrt{a}x} & (-\infty;0) \\ {\scriptstyle w_a\circ \mathrm{id}_v(x)=ax}\downarrow & & \downarrow{\scriptstyle \mathrm{id}_u} \\ (0;\infty) & \xrightarrow{x\mapsto -x/\sqrt{a}} & (-\infty;0) \end{array}$$

Hence, the representation of p with respect to the charts $(V, w_a \circ \mathrm{id}_v)$ and (U, id_u) is the map $x \mapsto -x/\sqrt{a}$.

Thus p is C^k in both charts of the atlas $\mathsf{S}(w_a)$, and therefore it is a self-diffeomorphism of $\mathfrak{S}(a)$.

(3) Again if $a \neq 1$, then by Theorem 4 and Corollary 3 the above bijections reduce to the following ones

$$\xi^{\mathrm{fix}}\colon \mathrm{Diff}^{\mathrm{fix}}_{\mathfrak{S}(a),\mathfrak{S}(1/a)}(\mathbb{L}) \to \mathrm{Diff}^k(\mathbb{R},0) \cap w_{1/a}\,\mathrm{Diff}^k(\mathbb{R},0)w_a^{-1} = \mathcal{E}^{k,-}(\mathbb{R},0),$$
$$\xi^{\mathrm{ex}}\colon \mathrm{Diff}^{\mathrm{ex}}_{\mathfrak{S}(a),\mathfrak{S}(1/a)}(\mathbb{L}) \to \mathrm{Diff}^k(\mathbb{R},0) \cap w_{1/a}\,\mathrm{Diff}^k(\mathbb{R},0)w_a = \mathcal{E}^{k,+}(\mathbb{R},0),$$

which send preserving (resp. reversing) orientation diffeomorphisms of $\mathbb{L}$ to diffeomorphisms of $\mathbb{R}$ of the same type.

Hence $\mathrm{Diff}^{+,\mathrm{fix}}_{\mathfrak{S}(a),\mathfrak{S}(1/a)}(\mathbb{L}) = \mathrm{Diff}^{-,\mathrm{ex}}_{\mathfrak{S}(a),\mathfrak{S}(1/a)}(\mathbb{L}) = \varnothing$, and

$$\mathrm{Diff}_{\mathfrak{S}(a),\mathfrak{S}(1/a)}(\mathbb{L}) = \mathrm{Diff}^{-,\mathrm{fix}}_{\mathfrak{S}(a),\mathfrak{S}(1/a)}(\mathbb{L}) \sqcup \mathrm{Diff}^{+,\mathrm{ex}}_{\mathfrak{S}(a),\mathfrak{S}(1/a)}(\mathbb{L}).$$

The latter statement on bijections of those sets with $\mathrm{Diff}^{+,\mathrm{fix}}_{\mathfrak{S}(a)}(\mathbb{L})$ follows from Lemma 10. This completes Theorem 2.

Acknowledgements This work was supported by grants from the Simons Foundation (1030291, 1290607, S.I.M. and L.M.V.).

References

1. L. Chen, K. Mann, Structure theorems for actions of homeomorphism groups. Duke Math. J. **172**(5), 915–962 (2023). https://doi.org/10.1215/00127094-2022-0019
2. S.K. Donaldson, An application of gauge theory to four-dimensional topology. J. Differ. Geom. **18**(2), 279–315 (1983). http://projecteuclid.org/euclid.jdg/1214437665
3. S. Ferry, Geometric topology notes. https://sites.math.rutgers.edu/~sferry/ps/geotop.pdf
4. J.S. Frame, The double cosets of a finite group. Bull. Am. Math. Soc. **47**, 458–467 (1941). https://doi.org/10.1090/S0002-9904-1941-07479-3
5. D.S. Freed, K.K. Uhlenbeck, *Instantons and Four-Manifolds*. Mathematical Sciences Research Institute Publications, vol. 1, 2nd edn. (Springer, New York, 1991). https://doi.org/10.1007/978-1-4613-9703-8
6. M.H. Freedman, The topology of four-dimensional manifolds. J. Differ. Geom. **17**(3), 357–453 (1982). http://projecteuclid.org/euclid.jdg/1214437136
7. D. Gauld. *Non-metrisable Manifolds* (Springer, Singapore, 2014). https://doi.org/10.1007/978-981-287-257-9
8. R.E. Gompf, An infinite set of exotic $\mathbf{R}^4$'s. J. Differ. Geom. **21**(2), 283–300 (1985). http://projecteuclid.org/euclid.jdg/1214439566

9. A. Haefliger, G. Reeb, Variétés (non séparées) à une dimension et structures feuilletées du plan. Enseignement Math. **2**(3), 107–125 (1957)
10. P. Hájíček, Causality in non-Hausdorff space-times. Comm. Math. Phys. **21**, 75–84 (1971). http://projecteuclid.org/euclid.cmp/1103857261
11. M. Hall, Jr., *The Theory of Groups* (The Macmillan Company, New York, 1959)
12. E. Hamilton, Double coset separability of abelian subgroups of hyperbolic n-orbifold groups. Proc. Am. Math. Soc. **144**(6), 2327–2336 (2016). https://doi.org/10.1090/proc/12904
13. E. Hamilton, H. Wilton, P.A. Zalesskii, Separability of double cosets and conjugacy classes in 3-manifold groups. J. Lond. Math. Soc. (2), **87**(1), 269–288 (2013). https://doi.org/10.1112/jlms/jds040
14. M. Heller, L. Pysiak, W. Sasin, Geometry of non-Hausdorff spaces and its significance for physics. J. Math. Phys. **52**(4), 043506 (2011). https://doi.org/10.1063/1.3574352
15. W. Sawin, Elementary proof of the uniqueness of smooth structures on $\mathbb{R}$. MathOverflow. https://mathoverflow.net/a/193832, https://mathoverflow.net/users/18060/willsawin
16. H. Kawabe, On the conjugation of local diffeomorphisms infinitely tangent to the identity, in *Foliations (Tokyo, 1983)*, Advanced Studies in Pure Mathematics (North-Holland, Amsterdam, 1985), pp. 461–481. https://doi.org/10.2969/aspm/00510461
17. M.A. Kervaire, A manifold which does not admit any differentiable structure. Comment. Math. Helv. **34**, 257–270 (1960). https://doi.org/10.1007/BF02565940
18. M.A. Kervaire, J.W. Milnor, Groups of homotopy spheres. I. Ann. of Math. **2**(77), 504–537 (1963). https://doi.org/10.2307/1970128
19. R.C. Kirby, L.C. Siebenmann, *Foundational Essays on Topological Manifolds, Smoothings, and Triangulations*. Annals of Mathematics Studies, No. 88 (Princeton University Press, Princeton, NJ; University of Tokyo Press, Tokyo, 1977). With notes by John Milnor and Michael Atiyah
20. A. Kowalik, I. Michalik, T. Rybicki, Reconstruction theorems for two remarkable groups of diffeomorphisms, in *Travaux mathématiques, vol. XVIII*, Travaux mathématiques, vol. 18 (Faculty of Science, Technology and Medicine - University of Luxembourg, Luxembourg, 2008), pp. 77–86
21. J. Kuan, Probability distributions of multi-species q-TAZRP and ASEP as double cosets of parabolic subgroups. Ann. Henri Poincaré **20**(4), 1149–1173 (2019). https://doi.org/10.1007/s00023-019-00761-y
22. J.M. Lee, *Introduction to Smooth Manifolds*. Graduate Texts in Mathematics, vol. 218, 2nd edn. (Springer, New York, 2013)
23. J.D.P. Meldrum, *Wreath Products of Groups and Semigroups*. Pitman Monographs and Surveys in Pure and Applied Mathematics, vol. 74 (Longman, Harlow, 1995)
24. J. Milnor, On manifolds homeomorphic to the 7-sphere. Ann. Math. **2**(64), 399–405 (1956). https://doi.org/10.2307/1969983
25. J. Munkres, Obstructions to the smoothing of piecewise-differentiable homeomorphisms. Ann. Math. **2**(72), 521–554 (1960). https://doi.org/10.2307/1970228
26. P.J. Nyikos, Various smoothings of the long line and their tangent bundles. Adv. Math. **93**(2), 129–213 (1992). https://doi.org/10.1016/0001-8708(92)90027-I
27. D. O'Connell, Non-Hausdorff manifolds via adjunction spaces. Topol. Appl. **326**, Paper No. 108388, 18 (2023). https://doi.org/10.1016/j.topol.2022.108388
28. D. O'Connell, Vector bundles over non-Hausdorff manifolds. Topol. Its Appl. 108982, (2024). https://doi.org/10.1016/j.topol.2024.108982
29. M. Rubin, On the reconstruction of topological spaces from their groups of homeomorphisms. Trans. Am. Math. Soc. **312**(2), 487–538 (1989). https://doi.org/10.2307/2001000
30. M. Rubin, T. Rybicki, Isomorphisms between groups of equivariant homeomorphisms of G-manifolds with one orbit type. Topol. Appl. **159**(12), 2899–2908 (2012). https://doi.org/10.1016/j.topol.2012.05.030
31. G.M. Seitz, Double cosets in algebraic groups, in *Algebraic Groups and Their Representations*, vol. 517 (Cambridge, 1997); NATO Advanced Science Institutes Series C: Mathematical and Physical Sciences (Kluwer Academic Publishers, Dordrecht, 1998), pp. 241–257

32. F. Sergeraert, Feuilletages et difféomorphismes infiniment tangents à l'identité. Invent. Math. **39**(3), 253–275 (1977). https://doi.org/10.1007/BF01402976
33. M. Spivak, *A Comprehensive Introduction to Differential Geometry*, vol. I, 2nd edn. (Publish or Perish Inc., Wilmington, DE, 1979)
34. J. Stallings, The piecewise-linear structure of Euclidean space. Proc. Cambridge Philos. Soc. **58**, 481–488 (1962)
35. S. Sternberg, Local C^n transformations of the real line. Duke Math. J. **24**, 97–102 (1957). http://projecteuclid.org/euclid.dmj/1077467216
36. F. Takens, Normal forms for certain singularities of vectorfields. Ann. Inst. Fourier (Grenoble) **23**(2), 163–195 (1973). http://www.numdam.org/item?id=AIF_1973__23_2_163_0
37. F. Takens, Characterization of a differentiable structure by its group of diffeomorphisms. Bol. Soc. Brasil. Mat. **10**(1), 17–25 (1979). https://doi.org/10.1007/BF02588337
38. C.H. Taubes, Gauge theory on asymptotically periodic 4-manifolds. J. Differ. Geom. **25**(3), 363–430 (1987). http://projecteuclid.org/euclid.jdg/1214440981
39. J.H.C. Whitehead, Manifolds with transverse fields in Euclidean space. Ann. Math. **2**(73), 154–212 (1961). https://doi.org/10.2307/1970286
40. J.V. Whittaker, On isomorphic groups and homeomorphic spaces. Ann. Math. **2**(78), 74–91 (1963). https://doi.org/10.2307/1970503

Generalized Geometry of 2D Incompressible Fluid Flows

Radek Suchánek

Abstract We describe a family of generalized almost structures associated with a Monge-Ampère equation for a stream function of a 2D incompressible fluid flow. Using an indefinite metric field constructed from a pair of 2-forms related to the Monge-Ampère equation, we show the existence of generalized metric compatible structures in our family of generalized structures. Integrability of isotropic structures on the level of Dirac structures and differential forms is discussed.

1 Introduction

In [4], Crainic described conditions on tensor fields to define a generic generalized complex structure [18, 19], and showed how Hitchin pairs of differential forms are connected with these structures. B. Banos showed in [12] that a certain class of non-linear second-order PDEs called Monge-Ampère equations gives rise to Hitchin pairs and the corresponding generalized complex geometry and generalized Calabi-Yau geometry [15, 17, 19]. The notion of integrability then translates to a certain local equivalence problem of Monge-Ampère equations and Jacobi system of nonlinear first-order PDEs with special non-linearity [15, 30]. Motivated by these results, we focus on a different approach toward the construction of generalized geometries and show that the family of generalized geometries described by Banos can be extended when dealing with incompressible fluids in 2D. In particular, we prove the existence of generalized Kähler structure and generalized chiral structures, as well as generalized hyper-complex and generalized hyper-para-complex triples. Applications of these structures in the context of twisted supersymmetric nonlinear sigma models were studied in [16].

The idea behind searching for geometric structures connected with Monge-Ampère equations is two-fold. Firstly, if one is interested in concrete examples of generalized structures with specific properties, Monge-Ampère theory provides numerous examples of geometric objects from which the generalized structures can be

R. Suchánek (✉)
Department of Mathematics and Statistics, Masaryk University, Brno, Czech Republic
e-mail: suchanek@math.muni.cz

M. Ulan and N. C. Combe (eds.), *Geometric Methods in Physical Systems: From Differentiable Structures to Applications*, Tutorials, Schools, and Workshops in the Mathematical Sciences, https://doi.org/10.1007/978-3-032-00399-7_2

constructed [2, 15, 30]. Secondly, investigating the geometry of equations may lead to a better understanding and possible improvement of certain models. An example of the usefulness of this approach is provided in [7, 9, 11, 14, 27, 30, 31].

2 Monge-Ampère Theory

All our manifolds will be real and smooth. Let $\mathcal{B}$ be (a real, smooth) two-dimensional manifold. Denote by x, y (local) coordinates on $\mathcal{B}$. In the following definition, we will describe a subclass of nonlinear second-order PDEs, with the non-linearities given by minors and the determinant of the Hessian matrix.

Definition 2.1 *Monge-Ampère equation on* $\mathcal{B}$ is the following nonlinear second-order PDE

$$Af_{xx} + 2Bf_{xy} + Cf_{yy} + D\left(f_{xx}f_{yy} - {f_{xy}}^2\right) + E = 0\,, \tag{1}$$

where $f_{xy} := \frac{\partial^2 f}{\partial x \partial y}$, and A, B, C, D, E are smooth functions, which depend on x, y, f_x, f_y.

Example 2.1
We provide a few standard examples of M-A equations. From (1) we see that the 2D Laplace equation $f_{xx} + f_{yy} = 0$, as well as the 2D wave equation $f_{xx} - f_{yy} = 0$, are constant coefficients M-A equations. These are particularly interesting in view of Proposition 2.1. Next examples are $\det \mathrm{Hess}(f) = 1$ and $\det \mathrm{Hess}(f) = -1$, which will play a key role in this paper, as they arise from the Navier-Stokes equations for the incompressible 2D fluid flows. Moreover, by virtue of Proposition 2.1, these equations are equivalent with the 2D Laplace and wave equations via a symplectic transformation. Let us end this short overview of M-A equations with the von Karman equation $f_x f_{xx} - f_{yy} = 0$. We will see in Example 2.3 that the von Karman equation has, in a certain sense, very different behavior than the previous examples.

Let $T^*\mathcal{B} \xrightarrow{\pi} \mathcal{B}$ be the cotangent bundle over $\mathcal{B}$. Throughout this paper, Ω will always denote the canonical symplectic form on the cotangent bundle. In the following, we will show how M-A equations can be encoded by a certain pair of differential 2-forms on $T^*\mathcal{B}$, called *M-A structure*. We start with the following observation. In the Darboux coordinates x, y, p, q (x, y are the base coordinates on $\mathcal{B}$), the canonical symplectic form writes as

$$\Omega = \mathrm{d}x \wedge \mathrm{d}p + \mathrm{d}y \wedge \mathrm{d}q\,. \tag{2}$$

Let α be a 2-form given by

$$\begin{aligned}\alpha = &A\mathrm{d}p \wedge \mathrm{d}y + B(\mathrm{d}x \wedge \mathrm{d}p - \mathrm{d}y \wedge \mathrm{d}q) \\ &+ C\mathrm{d}x \wedge \mathrm{d}q + D\mathrm{d}p \wedge \mathrm{d}q + E\mathrm{d}x \wedge \mathrm{d}y\,,\end{aligned} \tag{3}$$

where the coefficients A, B, C, D, E are smooth functions on $T^*\mathcal{B}$. Then $\alpha \wedge \Omega = 0$ holds for all possible choices of the coefficients.

Definition 2.2 ([1, 2]) A pair $(\Omega, \alpha) \in \Omega^2(T^*\mathcal{B}) \times \Omega^2(T^*\mathcal{B})$ is called *a 2D Monge-Ampère structure over* $T^*\mathcal{B}$ (or just *M-A structure*), if Ω is a symplectic form and the *effectivity condition* $\alpha \wedge \Omega = 0$ is satisfied.

Remark 2.1 Note that the coefficients $A, B, \ldots, E$ in (3) are functions on $T^*\mathcal{B}$, while the coefficients $A, B, \ldots, E$ given in (1) denoted functions on $\mathcal{B}$. We often denote by the same symbol a function $H \in C^\infty(T^*\mathcal{B})$, which depends on x, y, p, q, and its restriction to the graph of $\mathrm{d}f$, which depends on x, y, f_x, f_y. That is, we write $H = H(x, y, p, q)$, as well as $H = H(x, y, f_x, f_y) \equiv (\mathrm{d}f)^*H$. Similarly, we write $F = F(x, y) \in C^\infty\mathcal{B}$ as well as $F = F(x, y, 0, 0) \in C^\infty(T^*\mathcal{B})$. The domain of a function under consideration will always be clear from the context or stated explicitly.

2.1 *From M-A Structures to M-A Equations*

We will now describe the link between M-A structures and M-A equations [6, 20]. This will justify the nomenclature of Definition 2.2 and also explain the seemingly unnatural factor of 2 in the second summand of (1). Let $f : \mathcal{B} \to \mathbb{R}$ be a smooth function. The differential of f determines a section, $\mathrm{d}f : \mathcal{B} \to T^*\mathcal{B}$, given by $\mathrm{d}f(x) := \mathrm{d}_x f$, and we can pullback α onto $\mathcal{B}$ to obtain a top form $(\mathrm{d}f)^*\alpha \in \Omega^2(\mathcal{B})$. Then the equation

$$(\mathrm{d}f)^*\alpha = 0 \tag{4}$$

defines a nonlinear second-order PDE for f. If α is given by (3), then the Eq. (4) corresponds exactly to (1). For a detailed exposition on the theory of Monge-Ampère equations and their applications, particularly in dimensions two and three, see [2]. For further details about applications of Monge-Ampère theory, for example, in the theory of complex differential equations, see [13], theoretical meteorology, and incompressible fluid theory, see [7, 9–11, 14].

Example 2.2
Let $\alpha = -\mathrm{d}x \wedge \mathrm{d}q + \mathrm{d}y \wedge \mathrm{d}p$. Then

$$(\mathrm{d}f)^*\alpha = -\mathrm{d}x \wedge \mathrm{d}(f_y) + \mathrm{d}y \wedge \mathrm{d}(f_x) = (-f_{yy} - f_{xx})\mathrm{d}x \wedge \mathrm{d}y.$$

Hence $(\mathrm{d}f)^*\alpha = 0$ amounts to the 2D Laplace equation $f_{yy} + f_{xx} = 0$ and (Ω, α) is the corresponding M-A structure. Now let $\alpha = p\mathrm{d}p \wedge dy + \mathrm{d}x \wedge \mathrm{d}q$. Then $(\mathrm{d}f)^*\alpha = 0$ describes the von Karman equation $f_x f_{xx} - f_{yy} = 0$.

Definition 2.3 ([2]) *Pfaffian* of a M-A structure, $\mathrm{Pf}(\alpha)$, is defined by

$$\alpha \wedge \alpha = \mathrm{Pf}(\alpha)\Omega \wedge \Omega. \tag{5}$$

M-A structure is called *non-degenerate*, if the Pfaffian is nowhere-vanishing. If $\mathrm{Pf}(\alpha) > 0$, then she structure is called *elliptic*, and if $\mathrm{Pf}(\alpha) < 0$, then it is called *hyperbolic*. We call a non-degenerate structure *normalized*, if $|\,\mathrm{Pf}(\alpha)| = 1$.

For a general 2D M-A structure, the Pfaffian in canonical coordinates writes

$$\mathrm{Pf}(\alpha) = -B^2 + AC - DE. \tag{6}$$

Example 2.3
Let (Ω, α) be the M-A structure described in Example 2.2, i.e., structure for 2D Laplace equation. Then $\mathrm{Pf}(\alpha) = 1$. Now consider the M-A structure for the von Karman equation $f_x f_{xx} - f_{yy} = 0$. Then $\mathrm{Pf}(\alpha) = p$. In Sect. 2.2, we will see that the 2D Laplace equation gives rise to an integrable complex structure, while the von Karman equation does not.

Remark 2.2 In the modeling of stably stratified geophysical flows, the Pfaffian is related with Rellich's parameter [25, 26, 30].

A non-degenerate M-A structure can be normalized $(\Omega, \alpha) \mapsto (\Omega, n(\alpha))$ by $n(\alpha) := |\,\mathrm{Pf}(\alpha)|^{-\frac{1}{2}}\alpha$. Indeed, the Pfaffian of $n(\alpha)$ satisfies $|\,\mathrm{Pf}\,(n(\alpha))\,| = 1$, since

$$n(\alpha) \wedge n(\alpha) = |\,\mathrm{Pf}(\alpha)|^{-1}\alpha \wedge \alpha = \frac{\mathrm{Pf}(\alpha)}{|\,\mathrm{Pf}(\alpha)|}\Omega \wedge \Omega\,,$$

which implies

$$\mathrm{Pf}\,(n(\alpha)) = \mathrm{sgn}\,\mathrm{Pf}(\alpha). \tag{7}$$

A non-degenerate M-A structure (Ω, α) and its normalization $(\Omega, n(\alpha))$ correspond to the same M-A equation, since

$$(\mathrm{d}f)^* n(\alpha) = |\operatorname{Pf}(\alpha)|^{-\frac{1}{2}}\Big|_{\operatorname{Im} \mathrm{d}f} (\mathrm{d}f)^*\alpha = 0.$$

Remark 2.3 Notice that we could have rescaled α with an arbitrary non-vanishing function, which would result in a new M-A structure $(\Omega, \tilde{\alpha})$. Then, by the same argument as for the normalization, $(\Omega, \tilde{\alpha})$ defines the same M-A equation as (Ω, α). The main reason we choose to work with normalized structures is that some of the generalized geometries constructed from α are not invariant with respect to the rescaling. Hence, the normalization condition provides a consistent choice of the representative in the class $[\alpha]$, where $\tilde{\alpha} \in [\alpha]$, if and only if $\tilde{\alpha} = e^h \alpha$ for some function h.

2.2 Endomorphisms and Symmetric Form

In the following paragraphs, we want to describe how every non-degenerate M-A structure defines a field of endomorphisms, which square to either $\mathrm{Id}_{T(T^*\mathcal{B})}$, or $-\mathrm{Id}_{T(T^*\mathcal{B})}$. Before we do so, we need to fix some notation.

Notation. Tensor $\sigma \in \Gamma(T^*\mathcal{B} \otimes T^*\mathcal{B})$ can be identified with a $C^\infty(\mathcal{B})$-linear map $\sigma_\#\colon \Gamma(T\mathcal{B}) \to \Gamma(T^*\mathcal{B})$ defined by $\sigma_\#(X) := X \lrcorner\, \sigma$. Similarly, if $\tau \in \Gamma(T\mathcal{B} \otimes T\mathcal{B})$, then we denote by $\tau^\#$ the linear map $\tau^\#\colon \Gamma(T^*\mathcal{B}) \to \Gamma(T\mathcal{B})$, where $\tau^\#(\xi) := \xi \lrcorner\, \tau$. Now consider a non-degenerate 2-form $\alpha \in \Omega^2(T^*\mathcal{B})$, which means that $\alpha_\#\colon T\mathcal{B} \to T^*\mathcal{B}$ is an isomorphism. Using the inverse $(\alpha_\#)^{-1}\colon T^*\mathcal{B} \to T\mathcal{B}$, we define the bivector $\pi_\alpha \in \Gamma\left(\Lambda^2 T(T^*\mathcal{B})\right)$ by

$$(\pi_\alpha)^\# := (\alpha_\#)^{-1}. \tag{8}$$

For example, if $\Omega = \mathrm{d}x \wedge \mathrm{d}p + \mathrm{d}y \wedge \mathrm{d}q$ is the canonical symplectic form, then $\pi_\Omega = \partial_x \wedge \partial_p + \partial_y \wedge \partial_q$ is the corresponding bivector, where $\partial_x := \frac{\partial}{\partial_x}$ is the x-coordinate vector field (and similarly for the other fields ∂_i).

When working with matrices, we will use the underline notation to make the distinction between a morphism and its matrix representation. For example, if $\rho \in \operatorname{End}(T(T^*\mathcal{B}))$, then the corresponding matrix will be denoted $\underline{\rho}$ (which will always be understood with respect to the canonical coordinates of Ω). Also, we will be omitting the # symbol when dealing with the morphisms derived from 2-forms (and 2-vectors), e.g., the matrix of $\alpha_\#$ will be denoted simply $\underline{\alpha}$. Similarly, the matrix of $(\pi_\alpha)^\#$ will be denoted simply by $\underline{\alpha}^{-1}$ (see defining Eq. (8)). When dealing with generalized structures, we write $\mathbb{J}$ for a coordinate-free description, as well as for the coordinate description via matrices. Nevertheless, there is no space for confusion, since a block description of generalized structure either contains coordinate-free objects or their matrices, and the distinction between the two is clear in our notation.

Almost complex and almost product structure. Let $\rho \in \operatorname{End}(T(T^*\mathcal{B}))$ be an endomorphism defined by

$$\rho := |\operatorname{Pf}(\alpha)|^{-\frac{1}{2}} \pi_{\Omega}^{\#} \circ \alpha_{\#}. \tag{9}$$

If (Ω, α) is elliptic, then $\rho^2 = -\operatorname{Id}_{T(T^*\mathcal{B})}$, if (Ω, α) is hyperbolic, then $\rho^2 = \operatorname{Id}_{T(T^*\mathcal{B})}$. The fact that ρ is either an almost complex structure on $T^*\mathcal{B}$ (if $\operatorname{Pf}(\alpha) > 0$), or an almost product structure (if $\operatorname{Pf}(\alpha) < 0$), was proven in [20]. In the canonical coordinates,

$$\underline{\rho} = |\operatorname{Pf}(\alpha)|^{-\frac{1}{2}} \begin{pmatrix} B & -A & 0 & -D \\ C & -B & D & 0 \\ 0 & E & B & C \\ -E & 0 & -A & -B \end{pmatrix} \tag{10}$$

and, as expected, it follows that

$$\underline{\rho}^2 = \frac{-\operatorname{Pf}(\alpha)}{|\operatorname{Pf}(\alpha)|} \operatorname{Id}_{T(T^*\mathcal{B})} = -\operatorname{sgn} \operatorname{Pf}(\alpha) \operatorname{Id}_{T(T^*\mathcal{B})}.$$

Notice that ρ is invariant of the normalization, since (7) yields

$$\operatorname{sgn} \operatorname{Pf}(n(\alpha)) = \operatorname{sgn}^2 \operatorname{Pf}(\alpha) = \operatorname{sgn} \operatorname{Pf}(\alpha).$$

Integrability. V. Lychagin and V. Rubtsov showed in [20] that there is a direct link between local equivalence of M-A equations and integrability of the ρ structure derived from the corresponding M-A structure (Ω, α). Moreover, the integrability condition can be expressed as a certain closedness condition.

Proposition 2.1 (Lychagin-Rubtsov [20]) *A 2D symplectic M-A equation $(df)^*\alpha = 0$ can be locally transformed via a symplectic transformation to either the Laplace equation $\Delta f = 0$, or the wave equation $\Box f = 0$, if and only if the ρ structure is integrable, which is equivalent to $\frac{\alpha}{\sqrt{|\operatorname{Pf}(\alpha)|}}$ being closed.*

Symmetric bilinear form. Every M-A structure defines the following symmetric bilinear form on $T^*\mathcal{B}$ [2, 11]

$$g(X, Y) := \frac{2(X \lrcorner \alpha \wedge Y \lrcorner \Omega + Y \lrcorner \alpha \wedge X \lrcorner \Omega) \wedge \pi^* \operatorname{vol}}{\Omega \wedge \Omega}, \tag{11}$$

where $X, Y \in \Gamma(T(T^*\mathcal{B}))$, $\lrcorner$ is the interior product, and $\pi^* \operatorname{vol} \in \Omega^2(T^*\mathcal{B})$ is the pullback of a locally chosen top form $\operatorname{vol} \in \Omega^2(\mathcal{B})$ along the cotangent bundle projection $\pi : T^*\mathcal{B} \to \mathcal{B}$. Since in our considerations the symplectic form is fixed, we see that the above symmetric field is parameterized by a 2-form, and the definition (11) amounts to a mapping $\Omega^2(T^*M) \to S^2(T^*\mathcal{B})$, given by $\alpha \mapsto g$. The matrix of g in canonical coordinates is

$$\underline{g} = \begin{pmatrix} 2C & -2B & D & 0 \\ -2B & 2A & 0 & D \\ D & 0 & 0 & 0 \\ 0 & D & 0 & 0 \end{pmatrix}, \tag{12}$$

where A, B, C, D are the coefficients of α as in (3). Notice the independence of $\underline{g}$ on E. Moreover, it is clear that

$$\det \underline{g} \neq 0 \quad \Longleftrightarrow \quad D \neq 0.$$

Observe that the coordinate vector fields $\frac{\partial}{\partial p}, \frac{\partial}{\partial q}$ span a 2-dimensional totally isotropic subspace of $T(T^*\mathcal{B})$. Since rank $T(T^*\mathcal{B}) = 4$, it follows that g has signature $(2, 2)$. Applications of the symmetric form (11) in theoretical meteorology and the corresponding theory of fluid flows were discussed in [9, 11, 20]. Further information about the properties of g, as well as of the corresponding pullback metric $(df)^*g$, in the context of Ricci-flatness condition and Hessian structures was discussed in our previous work [23].

3 Incompressible Fluid Flows in 2D

The motivation for us to investigate the stream equation arising from the kinematics of incompressible 2D fluid flows are the works of Banos, Delahaies, Roubtsov, and Roulstone [7, 9, 10, 14]. In these papers, it was demonstrated that the modern differential geometric approach offers a new framework for understanding and studying semi-geostrophic theory and related models. In particular, the authors have shown the possibility of modeling coherent structures in large-scale atmosphere-ocean flows via a hierarchy of approximations to the Navier-Stokes equations and the related incompressible fluid flow models.

3.1 Kinematics of Incompressible Fluids

The starting point is the system

$$\partial_t \mathbf{v} + (\mathbf{v} \cdot \nabla)\mathbf{v} = \nu \Delta \mathbf{v} - \nabla P \,, \tag{13}$$

$$\nabla \cdot \mathbf{v} = 0 \tag{14}$$

where $\mathbf{v}$ is the *velocity field*, $\mathbf{v} = (a_t(x, y), b_t(x, y))$ (t is understood as a parameter), $\nu \in \mathbb{R}$ is the *viscosity* (which we assume constant), $P = P_t\,(x, y))$ is the *pressure field*, $\cdot$ is the dot product, $\nabla = \left(\partial_x, \partial_y\right)$, and the expression $\Delta \mathbf{v}$ refers to the vector Laplace operator $\Delta \mathbf{v} := \nabla\,(\nabla \cdot \mathbf{v}) - \nabla \times (\nabla \times \mathbf{v})$. Since we are working with $2D$ vectors, the vector product is understood as the composition

$$\mathbb{R}^2 \xrightarrow{\iota} \mathbb{R}^3 \xrightarrow{\nabla\times(-)} \mathbb{R}^3 \xrightarrow{\nabla\times(-)} \mathbb{R}^3 \xrightarrow{\mathrm{pr}} \mathbb{R}^2\,,$$

where $\iota(x, y) = (x, y, 0)$ and $\mathrm{pr}(x, y, z) = (x, y)$. The Navier-Stokes equation (13) and the *incompressibility constraint* (14) together (locally) describe the kinematics of 2D incompressible fluid flows.

3.2 M-A Equation for the Stream Function

Applying $\nabla \cdot (-)$ on (13) we obtain

$$\partial_t \nabla \cdot \mathbf{v} + \nabla \cdot ((\mathbf{v} \cdot \nabla)\mathbf{v}) = \nu \nabla \cdot (\Delta \mathbf{v}) - \Delta P. \tag{15}$$

Using the incompressibility constraint we have $\nabla \cdot \partial_t \mathbf{v} = \partial_t \nabla \cdot \mathbf{v} = 0$ and

$$\begin{aligned}\nabla \cdot ((\mathbf{v} \cdot \nabla)\mathbf{v}) &= (\partial_x a)^2 + \left(\partial_y b\right)^2 + 2\,\partial_y a\,\partial_x b\,,\\ &= -2\,\partial_x a\,\partial_y b + 2\,\partial_y a\,\partial_x b.\end{aligned}$$

Hence the left-hand side of the Eq. (15) can be written as

$$\nabla \cdot \partial_t \mathbf{v} + \nabla \cdot ((\mathbf{v} \cdot \nabla)\mathbf{v}) = -2 \det \begin{pmatrix} \partial_x a & \partial_y a \\ \partial_x b & \partial_y b \end{pmatrix}.$$

The term $\nu \nabla \cdot (\Delta \mathbf{v})$, appearing on the right-hand side of (15), vanish since

$$\nabla \cdot (\Delta \mathbf{v}) = \nabla \cdot [\nabla \times (\nabla \times \mathbf{v})] = \partial_{xxy} b - \partial_{xyy} a + \partial_{xyy} a - \partial_{xxy} b = 0.$$

The above computation leads to the following Poisson equation as a consequence of the system of Eqs. (13), (14)

$$\Delta P = 2 \det \begin{pmatrix} \partial_x a & \partial_y a \\ \partial_x b & \partial_y b \end{pmatrix}. \tag{16}$$

The fluid incompressibility Eq. (14) can be solved by $\mathbf{v} = \left(f_y, -f_x\right)$. This implies that (16) can be written as

$$\det \mathrm{Hess}(f) = \frac{1}{2}\Delta P. \tag{17}$$

Instead of viewing the above equation as a Poisson equation for P, one can instead interpret it as a Monge-Ampère equation for a function f of two variables, $f = f(x, y)$, called the *stream function*. This equation was studied by Roulston, Banos, Gibbon, and Roubtsov, in the context of semigeostrophic models [7, 29], where the relationship between sgn ΔP and the balance between the rate of strain and the enstrophy of the flow was discussed. In order to proceed further with our geometric approach, we assume $\Delta P \neq 0$.

3.3 M-A Geometry of Incompressible Fluid Flows

We now move toward the description of the M-A Eq. (17) via a pair of 2-forms (Ω, α) forming a normalized M-A structure. The corresponding pair of differential 2-forms is given by the canonical symplectic form (2) and

$$\alpha = \sqrt{\frac{2}{|\Delta P|}} \mathrm{d}p \wedge \mathrm{d}q - \frac{\Delta P}{\sqrt{|2\Delta P|}} \mathrm{d}x \wedge \mathrm{d}y \,, \tag{18}$$

where $\Delta P \neq 0$ is viewed as a function on $T^*\mathcal{B}$, which coincides with ΔP from Eq. (13) (see Remark 2.1). The pair clearly satisfies $\alpha \wedge \Omega = 0$, making it an M-A structure. This is in concordance with the general situation described in Sect. 2. Moreover,

$$(\mathrm{d}f)^*\alpha = \left(\sqrt{\frac{2}{|\Delta P|}}(f_{xx} f_{yy} - {f_{xy}}^2) - \frac{\Delta P}{\sqrt{|2\Delta P|}}\right) \mathrm{d}x \wedge \mathrm{d}y.$$

This implies that $(\mathrm{d}f)^*\alpha = 0$ amounts to $\det \mathrm{Hess}(f) = \frac{\Delta P}{2}$, and thus the above M-A structure corresponds to the M-A Eq. (17). From (18) we read the coefficients of α

$$A = 0 \,, B = 0 \,, C = 0 \,, D = \sqrt{\frac{2}{|\Delta P|}} \,, E = \frac{-\Delta P}{\sqrt{|2\Delta P|}}.$$

By direct computation, or with the help of (6), we see that the Pfaffian is

$$\mathrm{Pf}(\alpha) = \mathrm{sgn}\, \Delta P.$$

Hence the M-A structure is normalized in the sense of Definition 2.3. Consequently, the matrix of endomorphism ρ is

$$\underline{\rho} = |2\Delta P|^{-\frac{1}{2}} \begin{pmatrix} 0 & 0 & 0 & -2 \\ 0 & 0 & 2 & 0 \\ 0 & -\Delta P & 0 & 0 \\ \Delta P & 0 & 0 & 0 \end{pmatrix}. \tag{19}$$

Depending on the sgn ΔP, we obtain the following cases

$$\rho^2 = \begin{cases} -\operatorname{Id}_{T(T^*\mathcal{B})} & \text{for } \Delta P > 0\,, \\ \operatorname{Id}_{T(T^*\mathcal{B})} & \text{for } \Delta P < 0\,, \end{cases} \tag{20}$$

Let us proceed with the question of integrability. By virtue of Proposition 2.1, the integrability of ρ is equivalent to $\mathrm{d}\left(|\operatorname{Pf}(\alpha)|^{-\frac{1}{2}}\alpha\right) = 0$, where α is given by (18). Recall that the pressure is assumed to be a function of x, y only, $P = P(x, y)$, hence we have

$$\mathrm{d}\left(|\operatorname{Pf}(\alpha)|^{-\frac{1}{2}}\alpha\right) = \left(-\frac{\sqrt{2}}{2}|\Delta P|^{-\frac{3}{2}} \operatorname{sgn}\Delta P\right) \mathrm{d}\,(\Delta P) \wedge \mathrm{d}p \wedge \mathrm{d}q.$$

Since ΔP is a 0-form (hence cannot be closed), the last equation implies

$$\rho \text{ is integrable} \quad \Longleftrightarrow \quad \Delta P \text{ is constant.}$$

Indefinite metric field. To finish this section, we give the matrix description of the symmetric bilinear form (11) corresponding to (17). The matrix of g for a generic α is displayed in (12), which in our case becomes the following pseudo-metric field with split signature (2, 2)

$$\underline{g} = \sqrt{\frac{2}{|\Delta P|}} \begin{pmatrix} 0 & \mathbb{1} \\ \mathbb{1} & 0 \end{pmatrix}. \tag{21}$$

4 Generalized Geometry and M-A Theory

In this section we want to describe generalized structures associated with the equation for the stream function (17). For the sake of completeness, we recall some key definitions and objects from generalized geometry. To avoid any confusion, we formulate these notions for a general smooth manifold $\mathcal{M}$, and then choose $\mathcal{M} = T^*\mathcal{B}$ to investigate the generalized geometry of $\det \operatorname{Hess}(f) = 1$.

4.1 Generalized (Almost) Structures

A *generalized tangent bundle* over $\mathcal{M}$ is the vector bundle

$$\mathbb{T}\mathcal{M} := T\mathcal{M} \oplus T^*\mathcal{M} \xrightarrow{\ \pi\ } \mathcal{M}, \tag{22}$$

with the bundle projection π defined by the composition

$$\mathbb{T}\mathcal{M} \longrightarrow T\mathcal{M} \longrightarrow \mathcal{M},$$

where the left map is the projection on the first factor of the Whitney sum, and the second map is the tangent bundle projection on $\mathcal{M}$. The pairing between vector fields and 1-forms endows $\mathbb{T}\mathcal{M}$ with a non-degenerate, symmetric, bilinear form η

$$\eta\big((X, \xi), (Y, \zeta)\big) := \frac{1}{2}\left(\xi(Y) + \zeta(X)\right), \tag{23}$$

which defines on $\mathbb{T}\mathcal{M}$ a pseudo-Riemannian metric of signature (n, n). If we choose a coordinate system (q^{μ}) on $\mathcal{M}$, then the corresponding coordinate vector fields and 1-forms define a local basis $\big((\partial_{q^{\mu}}, 0), (0, dq^{\mu})\big)$ of $\mathbb{T}\mathcal{M}$. The matrix representation of η in this basis is

$$\eta = \begin{pmatrix} 0 & \mathbb{1} \\ \mathbb{1} & 0 \end{pmatrix}.$$

A generalized almost complex structure $\mathbb{J}$ *over* $\mathcal{M}$ is a bundle map $\mathbb{J}\colon \mathbb{T}\mathcal{M} \to \mathbb{T}\mathcal{M}$ such that $\mathbb{J}^2 = -\operatorname{Id}_{\mathbb{T}\mathcal{M}}$ and for all $(X, \eta), (Y, \theta) \in \mathbb{T}\mathcal{M}$

$$\langle \mathbb{J}(X, \eta), \mathbb{J}(Y, \theta) \rangle = \langle (X, \eta), (Y, \theta), \rangle \tag{24}$$

where η is the natural inner product (23) [18, 19].

Example 2.4
Let (Ω, α) be a non-degenerate M-A structure with α closed (i.e., a pair of symplectic structures). Consider a $(1, 1)$-tensor $A_{\alpha} := \pi^{\#}_{\Omega} \circ \alpha_{\#}$ (c.f. Definition 9). Then (Ω, A_{α}) defines *a Hitchin pair of 2-forms* in the sense of Crainic [4, 15] since

$$\Omega_{\#} A_{\alpha} = A^{*}_{\alpha} \Omega_{\#}.$$

Using the notion of Hitchin pairs, Banos [12] showed in that every 2D non-degenerate M-A structure satisfying for appropriate $\phi \in C^{\infty}(T^{*}\mathcal{B})$ the *divergence condition*

$$\mathrm{d}(\alpha + \phi\Omega) = 0\,,$$

yields an integrable generalized almost structure $\mathbb{J}_{\alpha}$ given as follows

$$\mathbb{J}_{\alpha} = \begin{pmatrix} A_{\alpha} & \pi^{\#}_{\Omega} \\ -\left(\Omega_{\#} + \Omega_{\#} A^{2}_{\alpha}\right) & -A^{*}_{\alpha} \end{pmatrix},$$

Table 1 Type of a generalized almost structure depending on (γ_1, γ_2)

(γ_1, γ_2)	$(1, 1)$	$(1, -1)$	$(-1, 1)$	$(-1, -1)$
Type of $\mathbb{J}$	GaP	GaPC	GaC	GaAC

The result described in Example 2.4 was a key motivation for our further investigations of other possibilities of constructing generalized (almost) geometries from Monge-Ampère equations and the corresponding Monge-Ampère structures. Starting with the M-A equation for the stream function (17) with specifically chosen ΔP, we will show that the family of generalized structures constructed with the help of M-A theory can be significantly enlarged. In order to state our result, we need to extend the notion of generalized (almost) complex structure by the following definition.

Definition 2.4 ([16]) *A generalized almost structure* $\mathbb{J}$ *over* $\mathcal{M}$ is a bundle map $\mathbb{J}\colon \mathbb{T}\mathcal{M} \to \mathbb{T}\mathcal{M}$ such that

$$\mathbb{J}^2 = \gamma_1 \operatorname{id}_{\mathbb{T}M} , \qquad \mathbb{J}^{\bullet}\eta = \gamma_2 \eta ,$$

where $\gamma_1, \gamma_2 \in \{-1, 1\}$ and $\mathbb{J}^{\bullet}\eta$ is understood in the sense of (24). Table 1 describes the four possible choices of constants γ_i. The abbreviations stand for *generalized almost product* (GaP), *generalized almost complex* (GaC), *generalized almost para-complex* (GaPC), and *generalized almost anti-complex* (GaAC) structure. A generalized structure is called *non-degenerate*, if its eigenbundles are isomorphic to $T\mathcal{M}$ (if $\mathbb{J}^2 = \operatorname{Id}_{\mathbb{T}\mathcal{M}}$), or to $T\mathcal{M} \otimes \mathbb{C}$ (if $\mathbb{J}^2 = -\operatorname{Id}_{\mathbb{T}\mathcal{M}}$).

Example 2.5
A family of examples of generalized almost structures comes from non-degenerate tensor fields of rank 2. Let $J \in \operatorname{Aut}(T\mathcal{M})$ be an almost complex structure, $\alpha \in \Omega^2(\mathcal{M})$ a non-degenerate 2-form, and $g \in S^2(\mathcal{M})$ a non-degenerate symmetric bilinear form. We have seen in the previous sections that M-A theory is a natural source of the above three objects. Let $\epsilon \in \{-1, 1\}$ and consider

$$\mathbb{J} = \begin{pmatrix} J & 0 \\ 0 & \epsilon J^* \end{pmatrix}, \qquad \mathbb{K} = \begin{pmatrix} 0 & \pi^{\#}_{\alpha} \\ \epsilon \alpha_{\#} & 0 \end{pmatrix}, \qquad \mathbb{G} = \begin{pmatrix} 0 & \pi^{\#}_{g} \\ \epsilon g_{\#} & 0 \end{pmatrix}. \tag{25}$$

Then $\mathbb{J}$ is a GaAC structure for $\epsilon = 1$, and a GaC structure for $\epsilon = -1$, while $\mathbb{K}$ is a GaPC structure for $\epsilon = 1$, and a GaC structure for $\epsilon = -1$. For $\mathbb{G}$, we have a GaP in the case $\epsilon = 1$ and a GaAC if $\epsilon = -1$.

We see that starting with non-degenerate rank 2 tensor fields, one can construct isotropic, as well as non-isotropic structures. We will end this example with the demonstration of the non-isotropy of the generalized almost product structure. Take $\mathbb{G}$ with $\epsilon = 1$. Then the ± 1-eigenbundles of $\mathbb{G}$ are

$$E_{\pm} = \{(X, \pm g_{\#}X) | X \in \Gamma(T\mathcal{M})\}.$$

Indeed, for $x_+ = (X, g_{\#}X) \in E_+$, and $y_- = (Y, -g_{\#}Y) \in E_-$, it holds $\mathbb{G}x_+ = (X, g_{\#}X) = x_+$, and $\mathbb{G}y = (-Y, g_{\#}Y) = -y_-$. Now suppose $x_+, y_+ \in E_+$. Then

$$\eta(x_+, y_+) = \frac{1}{2}((g_{\#}X)Y + (g_{\#}Y)X) = g(X, Y).$$

Obviously there are $X, Y \in \Gamma(T\mathcal{M})$ such that $g(X, Y) \neq 0$, thus E_+ is not totally isotropic, and hence $\mathbb{G}$ is a non-isotropic structure. Nonetheless, we can still speak about (non-)degeneracy of a non-isotropic structure.

The Courant bracket. The space of sections $\Gamma(\mathbb{T}\mathcal{M}) = \Gamma(T\mathcal{M}) \oplus \Gamma(T^*\mathcal{M})$ is equipped with the antisymmetric *Courant bracket* $[-, -]_C$ [5]

$$[(X, \xi), (Y, \zeta)]_C := \left([X, Y], \mathcal{L}_X\zeta - \mathcal{L}_Y\xi - \frac{1}{2}\mathrm{d}(X \lrcorner \zeta - Y \lrcorner \xi)\right), \tag{26}$$

where $[X, Y]$ is the Lie bracket and $\mathcal{L}$ is the Lie derivative. Note that the Courant bracket, which is an extension of the Lie bracket for vector fields, does not satisfy the Jacobi identity. More importantly for our considerations, the Courant bracket can be used to define integrability of *isotropic* structures on $\mathbb{T}\mathcal{M}$.

4.2 Isotropy, Dirac Structures, and Integrability

Every generalized almost structure comes together with the corresponding subbundles E_+, E_-, which are ± 1-eigenbundles if $\mathbb{J}^2 = \mathrm{id}_{\mathbb{T}M}$, and $\pm i$-eigenbundles if $\mathbb{J}^2 = -\mathrm{Id}_{\mathbb{T}M}$.

Definition 2.5 A subbundle $E \subset \mathbb{T}M$ is called *totally isotropic* (w.r.t. the inner product η), if for all $x, y \in E : \eta(x, y) = 0$. Totally isotropic E is called an *almost Dirac structure on M* if rank E = rank TM. A *Dirac structure on M* is an almost Dirac structure E such that $[E, E]_C \subset E$.

Remark 2.4 A Dirac structure on M can be equivalently defined as a totally isotropic subbundle $E \subset \mathbb{T}M$ of maximal rank, which is involutive with respect to the Courant bracket.

The four possible generalized almost structures determined by Definition 2.4 can be divided into two subsets depending on whether the eigenbundles $E_{\pm}$ are almost Dirac structures or not.

Definition 2.6 Let $\mathbb{J}$ be a generalized almost structure. If the eigenbundles $E_{\pm}$ are almost Dirac structures, then $\mathbb{J}$ is called *isotropic* (with respect to η). Otherwise $\mathbb{J}$ is called *non-isotropic*.

The involutivity condition of almost Dirac structures $E_{\pm}$ can serve as a definition of integrability only for isotropic $\mathbb{J}$ (see Remark 2.5 below).

Definition 2.7 An isotropic generalized almost structure $\mathbb{J} \in \mathrm{End}(\mathbb{T}M)$ is called *integrable* if the corresponding eigenbundles E_+, E_- are Dirac structures. An integrable generalized almost structure is called a *generalized structure*.

Let $\mathbb{J} \in \mathrm{End}(\mathbb{T}M)$ be an isotropic generalized almost structure. Then the torsion of $\mathbb{J}$ is defined for all $x, y \in \Gamma(\mathbb{T}M)$ by

$$N_{\mathbb{J}}(x, y) := [\mathbb{J}x, \mathbb{J}y]_C + \mathbb{J}^2[x, y]_C - \mathbb{J}\left([\mathbb{J}x, y]_C + [x, \mathbb{J}y]_C\right). \tag{27}$$

This $(1, 2)$-tensor $N_{\mathbb{J}} \colon \Gamma(\mathbb{T}M) \otimes \Gamma(\mathbb{T}M) \to \Gamma(\mathbb{T}M)$ is called (generalized) *Nijenhuis tensor*. An isotropic $\mathbb{A}$ is integrable if and only if the corresponding Nijenhuis tensor vanishes: $N_{\mathbb{A}}(x, y) = 0$ for all $x, y \in \Gamma(\mathbb{T}M)$ [4].

Remark 2.5 If a generalized almost structure $\mathbb{J}$ is non-isotropic, then the eigenbundles $E_{\pm}$ are not totally isotropic (see Example 2.5) and the Courant bracket is not well defined on them. Moreover, for non-isotropic structures, the torsion (27) is not a tensor. As a consequence, the notion of integrability, which is usually given either by the condition of vanishing Nijenhuis tensor, or by Definition 2.7, cannot be applied for non-isotropic structures. Nonetheless, there are other means of defining integrability of non-isotropic structures. For example, the authors of [16] defined *weak integrability* using the notion of generalized Bismut connection [17].

5 Generalized Geometry of 2D Incompressible Fluids

We have seen in Sect. 3.3 that, by virtue of the proposition 2.1, the endomorphism ρ is an integrable structure, if and only if ΔP is constant, nowhere-vanishing function. This also implies $\mathrm{d}\alpha = 0$. In the following proposition, we will focus on the case $\Delta P > 0$ and, for the convenience choose $\Delta P = 2$. In this situation, the equation for the stream function (17) becomes

$$\det \mathrm{Hess}(f) = 1. \tag{28}$$

The block matrices of ρ and $\alpha_{\#}$ are $\underline{\rho} = \begin{pmatrix} 0 & A \\ A & 0 \end{pmatrix}$ and $\underline{\alpha} = \begin{pmatrix} A & 0 \\ 0 & -A \end{pmatrix}$, where $A = \begin{pmatrix} 0 & -1 \\ 1 & 0 \end{pmatrix}$. We have $\underline{\rho}^2 = \underline{\alpha}^2 = -\mathbb{1}$, and $\underline{\rho}^T = -\underline{\rho}$. Notice that this relations change when $\Delta P = -2$. Of course, in all cases we have $\underline{\alpha}^T = -\underline{\alpha}$, since α is a 2-form.

Table 2 Generalized structure associated with $\det \mathrm{Hess}(f) = 1$

sgn ϵ_i	$\mathbb{J}_1$	$\mathbb{J}_2$	$\mathbb{J}_3$
$+$	GC	GC	GC
$-$	GaAC	GPC	GPC

Proposition 2.2 *Let (Ω, α) be the normalized M-A structure determined by Eq. (17) with $\Delta P = 2$. Let ρ be the corresponding endomorphism (9). Then the following $\mathbb{J}_i\colon \mathbb{T}\mathcal{B} \to \mathbb{T}\mathcal{B}$*

$$\mathbb{J}_1 = \begin{pmatrix} \rho & 0 \\ 0 & \epsilon_1 \rho^* \end{pmatrix}, \qquad \mathbb{J}_2 = \begin{pmatrix} 0 & \overline{\alpha} \\ \epsilon_2 \underline{\alpha} & 0 \end{pmatrix}, \qquad \mathbb{J}_3 = \begin{pmatrix} 0 & \overline{\Omega} \\ \epsilon_3 \underline{\Omega} & 0 \end{pmatrix}, \tag{29}$$

where $\epsilon_i \in \{-1, 1\}$, are generalized almost structures.

(1) If $\epsilon_1 = 1$, then $\mathbb{J}_1$ is isotropic and integrable. If $\epsilon_1 = -1$, then $\mathbb{J}_1$ is non-isotropic.
(2) The structures $\mathbb{J}_2, \mathbb{J}_3$ are isotropic and integrable.

The types of these structures, depending on ϵ_i are described in Table 2.

Proof The choice of the triple $(\mathbb{J}_i)$ corresponds to the Example (2.5) with respect to the M-A equation $\det \mathrm{Hess}(f) = 1$. This implies

$$(\mathbb{J}_1)^2 = -\mathbb{1}, \qquad (\mathbb{J}_2)^2 = \epsilon_2(-\mathbb{1}), \qquad (\mathbb{J}_3)^2 = \epsilon_3(-\mathbb{1}), \tag{30}$$

and

$$(\mathbb{J}_1)^*\eta = \epsilon_1 \eta, \qquad (\mathbb{J}_2)^*\eta = \epsilon_2 \eta, \qquad (\mathbb{J}_3)^*\eta = \epsilon_3 \eta, \tag{31}$$

where $(\mathbb{J}_i)^*\underline{\eta} = \mathbb{J}_i{}^T \underline{\eta} \mathbb{J}_i$. Due to our choice of ΔP, ρ is an integrable almost complex structure (a complex structure). If $\epsilon_1 = 1$, this implies that $\mathbb{J}_1$ is a generalized complex structure. If $\epsilon_1 = -1$, then $\mathbb{J}_1$ is a generalized almost anti-complex structure, which is non-isotropic. Furthermore, both α and Ω are closed. Thus for all choices of ϵ_2, ϵ_3, the structures $\mathbb{J}_2$ and $\mathbb{J}_3$ are integrable. Therefore $\mathbb{J}_2$ and $\mathbb{J}_3$ are (independently) either GC or GPC structures. □

5.1 Structures Generated by Anticommutative Triples

Let $\{\mathbb{J}_1, \mathbb{J}_2, \mathbb{J}_3\}$ be a triple of pair-wise anticommuting generalized almost structures, given by (29). The anticommutators are

$$\{\mathbb{J}_1, \mathbb{J}_2\} = \begin{pmatrix} 0 & -(\epsilon_1 - 1)\overline{\Omega} \\ \epsilon_2(\epsilon_1 - 1)\underline{\Omega} & 0 \end{pmatrix},$$

$$\{\mathbb{J}_2, \mathbb{J}_3\} = \begin{pmatrix} -(\epsilon_2 - \epsilon_3)\underline{\rho} & 0 \\ 0 & (\epsilon_2 - \epsilon_3)\underline{\rho} \end{pmatrix},$$

$$\{\mathbb{J}_1, \mathbb{J}_3\} = \begin{pmatrix} 0 & -(1 - \epsilon_1)\underline{\alpha} \\ \epsilon_3(1 - \epsilon_1)\underline{\alpha} & 0 \end{pmatrix},$$

and we observe

$$\{\mathbb{J}_i, \mathbb{J}_j\} = 0 \ \forall i, j \quad \Longleftrightarrow \quad \epsilon_1 = 1, \epsilon_2 = \epsilon_3.$$

This implies that there are exactly two possibilities in which the pair-wise anticommutativity occurs. In both cases $\mathbb{J}_1$ is a GC structure, and either $\mathbb{J}_2, \mathbb{J}_3$ are a pair of GC structures, or a pair of GPC structures. The first case consisting of a triple of GC structures has some further properties which are described in the Proposition 2.3 and the subsequent corollary. In both cases, any of the two structures generates the third one, and $\mathbb{J}_1$ is a fixed GC structure.

Proposition 2.3 *Let $\{\mathbb{J}_1, \mathbb{J}_2, \mathbb{J}_3\}$ be given by (29). Then the structures pair-wise anticommute,*

$$\{\mathbb{J}_i, \mathbb{J}_{i+1}\} = 0\ , \ i = 1, 2, 3\ ,$$

if, and only if, $\epsilon_1 = \epsilon_2 = \epsilon_3 = 1$, or, $\epsilon_1 = -\epsilon_2 = -\epsilon_3 = 1$.

(1) Let $\epsilon_1 = \epsilon_2 = \epsilon_3 = 1$. Then each $\mathbb{J}_i$ is a GC structure. Moreover, there is a 2-sphere of GaC structures in $\mathrm{span}_{\mathbb{R}}\{\mathbb{J}_1, \mathbb{J}_2, \mathbb{J}_3\}$.

(2) Let $\epsilon_1 = -\epsilon_2 = -\epsilon_3 = 1$. Then $\mathbb{J}_1$ is a GC structure and $\mathbb{J}_2, \mathbb{J}_3$ are GPC structures. Moreover, there is a one-sheeted hyperboloid of GaPC structures and a two-sheeted hyperboloid of GaC structures in $\mathrm{span}_{\mathbb{R}}\{\mathbb{J}_1, \mathbb{J}_2, \mathbb{J}_3\}$.

In both cases we have

$$\mathbb{J}_i \mathbb{J}_{i+1} = \mathbb{J}_{i+2} \mod 3\ , \ i = 1, 2, 3. \tag{32}$$

Proof Let $\mathbb{A}$ be an $\mathbb{R}$-linear combination of the almost generalized structures $\mathbb{J}_i$

$$\mathbb{A} = a_1\mathbb{J}_1 + a_2\mathbb{J}_2 + a_3\mathbb{J}_3. \tag{33}$$

From the mutual anticommutativity of the triple, we get

$$\mathbb{A}^2 = \left(a_1{}^2 + a_2{}^2\epsilon_2 + a_3{}^2\epsilon_3\right)(-\mathbb{1})\ ,$$
$$\mathbb{A}^*\eta = (a_1{}^2 + a_2{}^2\epsilon_2 + a_3{}^2\epsilon_3)\eta.$$

Suppose $\epsilon_1 = \epsilon_2 = \epsilon_3 = 1$. Then for any $(a_1, a_2, a_3) \in \mathbb{R}^3$ such that $a_1{}^2 + a_2{}^2 + a_3{}^2 = 1$ we have $\mathbb{A}^2 = -\mathbb{1}$ and $\mathbb{A}^*\eta = \eta$. Thus $\mathbb{A}$ is a GaC structure. Now suppose $\epsilon_1 = -\epsilon_2 = -\epsilon_3 = 1$. Then

$$\mathbb{A}^2 = \left(-{a_1}^2 + {a_2}^2 + {a_3}^2\right) \mathbb{1} , \tag{34}$$

$$\mathbb{A}^* \eta = \left({a_1}^2 - {a_2}^2 - {a_3}^2\right) \eta . \tag{35}$$

Thus for $-{a_1}^2 + {a_2}^2 + {a_3}^2 = 1$ we have $\mathbb{A}^2 = \mathbb{1}$ and $\mathbb{A}^*\eta = -\eta$, so $\mathbb{A}$ is a GaPC structure. Similarly, for $-{a_1}^2 + {a_2}^2 + {a_3}^2 = -1$ we have $\mathbb{A}^2 = -\mathbb{1}$ and $\mathbb{A}^*\eta = \eta$, hence $\mathbb{A}$ is a GaC structure. The remaining statements of the proposition were discussed before the proposition. □

Definition 2.8 ([21, 22]) *A generalized almost hyper-complex structure* is a triple $\{\mathbb{A}, \mathbb{B}, \mathbb{C}\}$ of GaC structures such that $\mathbb{A}\mathbb{B}\mathbb{C} = -\mathbb{1}$. *A generalized almost hyper-para-complex structure* is a triple $\{\mathbb{A}, \mathbb{B}, \mathbb{C}\}$, consisting of a GaC structure and a pair of GaPC structures such that $\mathbb{A}\mathbb{B}\mathbb{C} = \mathbb{1}$. The structures are integrable, if $\{\mathbb{A}, \mathbb{B}, \mathbb{C}\}$ are integrable. In this case we call the triple *generalized hyper-complex* and *generalized hyper-para-complex* structure, respectively.

Remark 2.6 In [22], the notion of a generalized hyper-complex structure is named *hyper-complex structure on a Courant algebroid* (in our case, the Courant algebroid is the generalized tangent bundle).

Proposition 2.4 *The triplet of pair-wise anticommuting generalized structures given by (29) satisfies*

$$\mathbb{J}_1\mathbb{J}_2\mathbb{J}_3 = \epsilon_3(-\mathbb{1}). \tag{36}$$

(1) If $\epsilon_3 = 1$, then $\{\mathbb{J}_1, \mathbb{J}_2, \mathbb{J}_3\}$ is a generalized hyper-complex structure on $\mathcal{B}$.
(2) If $\epsilon_3 = -1$, then $\{\mathbb{J}_1, \mathbb{J}_2, \mathbb{J}_3\}$ is a generalized hyper-para-complex structure on $\mathcal{B}$.

Proof From Proposition (2.3) we know, independently of the value of ϵ_3, that $\mathbb{J}_1\mathbb{J}_2 = \mathbb{J}_3$ and $\epsilon_1 = 1$, which means that all the three structures are integrable. Due to (30), we have $\mathbb{J}_1\mathbb{J}_2\mathbb{J}_3 = \epsilon_3(-\mathbb{1})$. If $\epsilon_3 = 1$, then $\epsilon_2 = 1$, and thus $\{\mathbb{J}_1, \mathbb{J}_2, \mathbb{J}_3\}$ is a triple of GC structures (cf. Table 2). Moreover, $\mathbb{J}_1\mathbb{J}_2\mathbb{J}_3 = -\mathbb{1}$, so the triple is a generalized hyper-complex structure. On the other hand, if $\epsilon_3 = -1$, then $\epsilon_2 = -1$, and thus $\{\mathbb{J}_1, \mathbb{J}_2, \mathbb{J}_3\}$ contains a GC structure and a pair of GPC structures (cf. Table 2). Moreover, $\mathbb{J}_1\mathbb{J}_2\mathbb{J}_3 = \mathbb{1}$, so the triple is a generalized hyper-para-complex structure. □

5.2 Generalized Metric Compatible Structures

We say that a generalized almost product structure $\mathbb{G} \in \text{End}\,(T\mathcal{M})$ is *a generalized (pseudo-)metric*, if the eigenbundles of $\mathbb{G}$ are isomorphic to $T\mathcal{M}$, that is, if $\mathbb{G}$ is non-degenerate.

Example 2.6
The main example of an indefinite generalized metric can be constructed from an indefinite metric (11) as $\mathbb{G} = \begin{pmatrix} 0 & \pi_g^\# \\ g_\# & 0 \end{pmatrix}$.

Definition 2.9 ([16]) *A generalized Kähler structure* is a commuting pair of a generalized metric and a GaC structure. *A generalized chiral structure* is a commuting pair of a generalized metric and a GaP structure.

Proposition 2.5 *The M-A equation* $\det \text{Hess}(f) = \frac{1}{2}\Delta P$ *defines a generalized Kähler structure for* $\Delta P = 2$*, and a generalized chiral structure for* $\Delta P = -2$.

Proof Let $\mathbb{G}$ and $\mathbb{J}$ be given by the following block matrices

$$\mathbb{G} = \begin{pmatrix} 0 & \underline{g} \\ \underline{g} & 0 \end{pmatrix}, \qquad \mathbb{J} = \begin{pmatrix} \underline{\rho} & 0 \\ 0 & \text{sgn}\,(\Delta P)\,\underline{\rho} \end{pmatrix},$$

where $\underline{g} = \begin{pmatrix} 0 & \mathbb{1} \\ \mathbb{1} & 0 \end{pmatrix}$ corresponds to (21), with $|\Delta P| = 2$, and ρ is given by (21)

$$\underline{\rho} = \begin{pmatrix} 0 & A \\ \text{sgn}\,(\Delta P)\,A & 0 \end{pmatrix},$$

where $A = \begin{pmatrix} 0 & -1 \\ 1 & 0 \end{pmatrix}$ Notice that $\mathbb{G}$ and $\mathbb{J}$ correspond to structures from (25) with $\epsilon = 1$, and where ρ is also allowed to be an almost product structure. Then

$$[\mathbb{G}, \mathbb{P}] = 0 \quad \Longleftrightarrow \quad \begin{cases} [\underline{g}, \underline{\rho}] = 0 & \text{for } \Delta P = 2 \\ \{\underline{g}, \underline{\rho}\} = 0 & \text{for } \Delta P = -2. \end{cases} \tag{37}$$

A direct computation shows that $[\underline{g}, \underline{\rho}] = 0$ is true for $|\Delta P| = 2$ (i.e., independently of the sign of Pfaffian). Thus $\mathbb{G}$ commutes with $\mathbb{J}$ for both values of ΔP. Moreover, $\underline{\rho}$ is a complex structure for $\Delta P = 2$ and a product structure for $\Delta P = -2$, which further implies that $\mathbb{J}$ is a GC structure for $\Delta P = 2$ and a GP structure for $\Delta P = -2$. Hence for $\Delta P = 2$, the pair $(\mathbb{G}, \mathbb{J})$ is a generalized Kähler structure, while for $\Delta P = -2$, the pair is a generalized chiral structure. □

5.3 Algebras Generated by Generalized Structures

There are multiple algebraic structures we can associate to a triple of generalized structures (29). The main point of the following paragraph is to discuss the possibility of studying M-A equations and M-A structures on the level of algebraic structures determined by their generalized geometries. The hypothesis which we want to investigate in our future work is whether the algebras defined below are invariant with respect to certain transformations. For example, one might study invariance with respect to symplectomorphisms, which provide a notion of equivalence between M-A equations (see Proposition 2.1).

Let us denote by $\mathcal{A}(\mathbb{J}_i, \epsilon_i)_{i=1}^3$ the free, associative, unital $\mathbb{R}$-algebra, given by the real vector space generated by the generalized almost structures $\{\mathbb{J}_i\}_{i=1}^3$ with fixed values $(\epsilon_i)_{i=1}^3$, and equipped with the product of matrices (or composition of operators, if we work coordinate-free)

$$\mathcal{A}(\mathbb{J}_i, \epsilon_i)_{i=1}^3 := \left(\mathbb{R}[\mathbb{J}_i]_{i=1}^3, (\epsilon_i)_{i=1}^3, \cdot\right). \tag{38}$$

We will omit the $\cdot$ symbol when writing product of elements from $\mathcal{A}(\mathbb{J}_i, \epsilon_i)_{i=1}^3$. We further consider two other bilinear operations on this algebra, the commutator and the anticommutator.

Jordan algebra. By a Jordan algebra we mean a commutative (possibly non-associative) algebra which satisfies $(AB)(AA) = A(B(AA))$ [28]. From every associative k-algebra[1] $(\mathcal{A}, \cdot)$ we can construct a Jordan k-algebra by substituting the algebra product with half of the anticommutator

$$A \bullet B := \frac{1}{2}\{A, B\}. \tag{39}$$

The resulting Jordan algebra is called a *special Jordan algebra*. Let us denote the special Jordan algebra constructed this way from $\mathcal{A}(\mathbb{J}_i, \epsilon_i)_{i=1}^3$ by $\mathrm{Jor}(\mathbb{J}_i, \epsilon_i)_{i=1}^3$, i.e.,

$$\mathrm{Jor}(\mathbb{J}_i, \epsilon_i)_{i=1}^3 := \left(\mathbb{R}[\mathbb{J}_i]_{i=1}^3, (\epsilon_i)_{i=1}^3, \bullet\right). \tag{40}$$

Lie algebra. We will denote the Lie algebra generated by the generalized structures $\{\mathbb{J}_i\}_{i=1}^3$ with fixed epsilons $(\epsilon_i)_{i=1}^3$ by $\mathrm{Lie}(\mathbb{J}_i, \epsilon_i)_{i=1}^3$, i.e.,

$$\mathrm{Lie}(\mathbb{J}_i, \epsilon_i)_{i=1}^3 := \left(\mathbb{R}[\mathbb{J}_i]_{i=1}^3, (\epsilon_i)_{i=1}^3, [-, -]\right). \tag{41}$$

Poisson algebra. We have two Poisson algebra structures at hand. One of them is coming from a Lie algebra, the other one from a Jordan algebra.

(1) Since the underlying algebra $\mathcal{A}(\mathbb{J}_i, \epsilon_i)_{i=1}^3$ is associative and the Lie bracket is given by the commutator, we automatically have a Poisson algebra at hand

$$\mathrm{Poiss}(\mathbb{J}_i, \epsilon_i)_{i=1}^3 := \left(\mathcal{A}(\mathbb{J}_i, \epsilon_i)_{i=1}^3, [-, -]\right). \tag{42}$$

Indeed, for every $\mathbb{A}, \mathbb{B}, \mathbb{C} \in \mathcal{A}(\mathbb{J}_i, \epsilon_i)_{i=1}^3$ we have

[1] Where k is a field of characteristic not equal to 2.

$$[\mathbb{A}, \mathbb{B}\mathbb{C}] = [\mathbb{A}, \mathbb{B}]\mathbb{C} + \mathbb{B}[\mathbb{A}, \mathbb{C}]. \tag{43}$$

(2) The second Poisson algebra is given by considering the (non-associative) Jordan algebra $\mathrm{Jor}(\mathbb{J}_i, \epsilon_i)_{i=1}^3$, equipped with the commutator

$$\mathrm{Poiss}(\mathbb{J}_i, \epsilon_i, \bullet)_{i=1}^3 := \left(\mathrm{Jor}(\mathbb{J}_i, \epsilon_i)_{i=1}^3, [-, -]\right). \tag{44}$$

Indeed, for every $\mathbb{A}, \mathbb{B}, \mathbb{C} \in \mathbb{R}[\mathbb{J}_i]_{i=1}^3$ we have

$$[\mathbb{A}, \mathbb{B} \bullet \mathbb{C}] = [\mathbb{A}, \mathbb{B}] \bullet \mathbb{C} + \mathbb{B} \bullet [\mathbb{A}, \mathbb{C}]. \tag{45}$$

Remark 2.7 In the second case of Poisson algebra coming from a Jordan algebra, one might wonder whether we have a Lie-Jordan algebra at hand, i.e., an algebraic structure which is both, a Lie algebra, a Jordan algebra, and the two bilinear operations are compatible in the following sense

$$[\mathbb{A}, \mathbb{B} \bullet \mathbb{C}] = [\mathbb{A}, \mathbb{B}] \bullet \mathbb{C} + \mathbb{B} \bullet [\mathbb{A}, \mathbb{C}] , \tag{46}$$

$$q^2[[\mathbb{A}, \mathbb{C}], \mathbb{B}] = (\mathbb{A} \bullet \mathbb{B}) \bullet \mathbb{C} - \mathbb{A} \bullet (\mathbb{B} \bullet \mathbb{C}) , \tag{47}$$

where $q \in \mathbb{R}$ (or $q \in k$, if we consider a different underlying field). As we have stated above, the first axiom is satisfied. *The second axiom is satisfied if, and only if* $q = \frac{\sqrt{-1}}{2}$.

Remark 2.8 Lychagin developed in [24] all the necessary definitions of algebraic structures, such as Lie algebras and Jordan algebras, defined for a system of two-dimensional nonlinear first-order partial differential equations with a specific type of non-linearity. A subclass of these systems is called *Jacobi systems*. In our future work, we will be interested in the extension of our algebraic setup to such systems, based on the results in [24].

6 Conclusion

With the help of Monge-Ampère theory [1, 2, 24], we have constructed a family of generalized structures associated to a Monge-Ampère equation (17), which was derived from the Navier-Stokes system for 2D incompressible fluid flows. Distinguishing between the hyperbolic and elliptic cases, we have shown that many of these generalized structures are integrable and give rise to further composite structures such as hyper-complex and hyper-para-complex. Introducing an indefinite generalized metric defined with the help of pseudo-metric (21), we have also found generalized metric compatible structures, namely generalized Kähler and generalized chiral structures.

Our future work will be focused on the study of other M-A equations and their relation to generalized geometry, with the aim to go into higher dimensions. Particularly interesting seems to be dimension three, as the situation there is much richer, yet the symplectic classification of differential forms is still possible. In this context, we want to further investigate the correspondences between M-A equations, families of generalized structures, and algebras constructed from these structures. We are especially curious which kind of equivalence notions naturally present in the M-A theory are able to survive when passing to either generalized structures, or various algebras we have linked with them. Yet another direction to investigate is the possibility of extending this framework to the Jacobi systems and other more general situations of first or second-order PDEs with special types of non-linearities, since a framework for doing so has already been developed [24].

Acknowledgements First of all I want to express my wholehearted gratitude to my beloved wife, son, and daughter for their continuous support and patience. This paper was initially written as a part of my Ph.D. research under the cotutelle agreement between the Masaryk University, Brno, Czech Republic, and the University of Angers, France, and finished during my 2023–2024 postdoc position at the Masaryk University, financed by the project GAČR GA22-00091S. I am grateful for the hospitality and support of these institutions during the research periods. I want to express my gratefulness to Volodya Rubtsov and Jan Slovák for their supervision over my research project and their overall support of my academic career. Preliminary results of this paper were reported at the Winter School and Workshop Wisla 22, a European Mathematical Society event organized by the Baltic Institute of Mathematics.

References

1. V. Lychagin, Contact geometry and non-linear second-order differential equations. Russ. Math. Surv. **34**, 149–180 (1979)
2. A. Kushner, V. Lychagin, V. Rubtsov, *Contact Geometry and Nonlinear Differential Equations* (Cambridge University Press, 2006)
3. I. Kolář, P. Michor, J. Slovák, *Natural Operations in Differential Geometry* (Springer, Berlin Heidelberg, 1993). https://www.emis.de/monographs/KSM/
4. M. Crainic, Generalized complex structures and Lie brackets. New Series. Bull. Braz. Math. Soc. **42**, 559–578 (2004)
5. T. Courant, Dirac manifolds. Trans. Am. Math. Soc. **319**, 631–661 (1990). http://www.jstor.org/stable/2001258
6. V. Lychagin, V. Rubtsov, I. Chekalov, A classification of Monge-Ampère equations. Ann. Scientifiques De L'École Normale Supérieure **26**(4), 281–308 (1993). http://www.numdam.org/articles/10.24033/asens.1673/
7. I. Roulstone, B. Banos, J. Gibbon, V. Roubtsov, Kähler geometry and Burgers' vortices. Proc. Ukr. Natl. Acad. Math. **16**, 303–321 (2009)
8. S. Amari, J. Armstrong, Curvature of Hessian manifolds. Differ. Geom. Its Appl. **33**, 1–12 (2014). https://www.sciencedirect.com/science/article/pii/S0926224514000023
9. B. Banos, V. Rubtsov, I. Roulstone, Monge–Ampère structures and the geometry of incompressible flows. J. Phys. A: Math. Theor. **49** (2016)
10. V. Rubtsov, I. Roulstone, Examples of quaternionic and Kähler structures in Hamiltonian models of nearly geostrophic flow. J. Phys. A: Math. Gen. **30**, L63–L68 (1997). https://doi.org/10.1088/0305-4470/30/4/004

11. V. Rubtsov, I. Roulstone, Holomorphic structures in hydrodynamical models of nearly geostrophic flow. Proc. R. Soc. Lond. A. **457**, 1519–1531 (2001)
12. B. Banos, Monge–Ampère equations and generalized complex geometry—the two-dimensional case. J. Geom. Phys. **57**, 841–853 (2007). https://www.sciencedirect.com/science/article/pii/S0393044006000726
13. B. Banos, Complex solutions of Monge–Ampère equations. J. Geom. Phys. **61**, 2187–2198 (2011). https://www.sciencedirect.com/science/article/pii/S0393044011001641
14. S. Delahaies, Complex and contact geometry in geophysical fluid dynamics (ProQuest LLC, 2008, 1)
15. Y. Kosmann-Schwarzbach, V. Rubtsov, Compatible structures on lie algebroids and Monge-Ampère operators. Acta Applicandae Mathematicae **109**, 101–135 (2010). https://doi.org/10.1007/s10440-009-9444-2
16. S. Hu, R. Moraru, D. Svoboda, Commuting Pairs, Generalized para-Kähler Geometry and Born Geometry (2019). arXiv:1909.04646
17. M. Gualtieri, Branes on Poisson varieties, in *The Many Facets of Geometry*, pp. 368–394 (2010)
18. N. Hitchin, Generalized Calabi–Yau manifolds. Q. J. Math. **54**, 281–308 (2003). https://doi.org/10.1093/qmath/hag025
19. M. Gualtieri, Generalized complex geometry. Ann. Math. **174**, 75–123 (2011). http://www.jstor.org/stable/23030560
20. V. Lychagin, V. Rubtsov, Local classification of Monge-Ampère differential equations. Sov. Math., Dokl. **28**, 328–332 (1983)
21. V. Cortés, L. David, Generalized connections, spinors, and integrability of generalized structures on courant algebroids (2021)
22. W. Hong, M. Stiénon, From hypercomplex to holomorphic symplectic structures. J. Geom. Phys. **96**, 187–203 (2015). https://www.sciencedirect.com/science/article/pii/S0393044015001308
23. S. Hronek, R. Suchánek, Pseudo-Riemannian and Hessian geometry related to Monge-Ampère structures. Archivum Mathematicum. **58**, 329–338 (2022)
24. V. Lychagin, Differential equations on two-dimensional manifolds. Izv. Vyssh. Uchebn. Zaved. Mat. **5**, 43–57 (1992)
25. D. Dritschel, Á. Viúdez, A balanced approach to modelling rotating stably stratified geophysical flows. J. Fluid Mech. **488**, 123–150 (2003)
26. Á. Viúdez, D. Dritschel, An explicit potential-vorticity-conserving approach to modelling nonlinear internal gravity waves. J. Fluid Mech. **458**, 75–101 (2002)
27. I. Roulstone, M. Sewell, The mathematical structure of theories of semigeostrophic type. Philos. Trans. R. Soc. London. Ser. A: Math., Phys. Eng. Sci. **355**, 2489–2517 (1997). https://royalsocietypublishing.org/doi/abs/10.1098/rsta.1997.0144
28. F. Ciaglia, J. Jost, L. Schwachhofer, What Lie algebras can tell us about Jordan algebras (2021)
29. I. Roulstone, B. Banos, J. Gibbon, V. Roubtsov, A geometric interpretation of coherent structures in Navier-Stokes flows. Proc. R. Soc. Lond. Ser. A Math. Phys. Eng. Sci. **465**, 2015–2021 (2009)
30. V. Rubtsov, *Geometry of Monge-Ampère structuress* (Birkhäuser/Springer, Cham, 2019)
31. M. McIntyre, I. Roulstone, Are there higher-accuracy analogues of semigeostrophic theory? (Cambridge University Press, 2002)

Nonlinear Differential Equations of Fluid Mechanics: Symmetries, Integrability, Singularities

Michael Roop

Abstract In this paper, we illustrate how the ideas from contact and symplectic geometry, as well as from the geometric theory of partial differential equations, can be applied to finding and investigating singular properties of solutions to fundamental equations of fluid dynamics. We give a geometric description of barotropic Euler's equations on curves and discuss their integrability both in the Jacobi-type systems setting and via differential invariants and quotient PDEs. New classes of multivalued solutions are found, and the description of caustics and shock wave fronts is given. The appearance of phase transitions along the flow in the case of a van der Waals gas is discussed. For barotropic Euler's equations in 2D and 3D, we find explicit multivalued solutions and discuss singularities of their projections to the space of independent variables. Finally, we address fully compressible Euler's equations by a similar approach.

1 Introduction

This paper is a continuation of a series of works [1–3] devoted to the investigation of singular properties of solutions to nonlinear PDEs of fluid dynamics. Here, we summarize and overview the results obtained in the above-mentioned papers, while presenting some new results on axially and spherically symmetric Euler's flows in two and three dimensions, as well as fully compressible Euler's flows.

A system of equations that describes the motion of a gas or a fluid consists of nonlinear partial differential equations representing the conservation of energy, momentum, and mass [4, 5]. The conservation of momentum may take various forms depending on assumptions and simplifications that are made on the basis of keeping the most important features of medium and the nature of the flow under consideration, while neglecting the others. For example, in the most general setting of viscous compressible flows it takes the form of the Navier-Stokes equation, in case of inviscid

M. Roop (✉)
Department of Mathematical Sciences, Chalmers University of Technology and University of Gothenburg, Gothenburg, Sweden
e-mail: michael.roop@chalmers.se

M. Ulan and N. C. Combe (eds.), *Geometric Methods in Physical Systems: From Differentiable Structures to Applications*, Tutorials, Schools, and Workshops in the Mathematical Sciences, https://doi.org/10.1007/978-3-032-00399-7_3

fluids it takes the form of the Euler equation, and for flows in porous media it is the Darcy law [6–10]. However, in any case the number of unknowns is not balanced by the number of equations, namely, some relations that represent thermodynamic properties of the medium called *state equations* [11] are to be added to make the system complete. This means that thermodynamics of the medium plays an important role in its motion.

It has been known since the nineteenth century [12] that geometric framework is natural in description of thermodynamic states. Recently, this approach has been revisited [13] and comprehensively reformulated in modern terms of contact and symplectic geometry that allowed to make use of theory of Lagrangian and Legendrian singularities developed by Arnold [14–16] to study *phase transitions*—jump-like transitions between different qualitatively different states of the medium (say, liquid and gas phases). On the other hand, equations that we study in fluid mechanics, and in particular in this paper, are nonlinear, which means that their solutions may have singularities and in general do not exist globally in time. In the context of non-stationary gas dynamics such solutions can be interpreted as shock waves. In the present paper we study these two types of singularities, phase transitions and shock waves, that constitute what we call *critical phenomena*. Similar effects in stationary fluid dynamics have previously been studied in works [17–20], and this paper concentrates on non-stationary fluid dynamics.

The first and probably the main hero of this paper is barotropic Euler's equations in one spatial dimension. Various critical properties of their solutions have always been of particular interest [21–23]. Our approach to studying and investigating singular properties of their solutions is essentially based on the geometric theory of PDEs [24–27]. This system is known to be of Jacobi type [28], which means that it can be represented by differential 2-forms on the 0-jet space. This approach originated from a seminal paper [29], which gave rise to numerous works, both theoretical and applied, in particular, [30–32], where singularities of solutions to incompressible hydrodynamics equations are discussed. It is worth also mentioning a series of works [33–36] on global solvability of barotropic Euler's equations that use the similar technique. Among other things, this approach makes it possible to generalize the classical solution to this system in a pure geometric way by saying that any integral manifold of these forms is considered to be a solution and allows one to work out the appearance of singularities staying in the smooth manifolds setting. Namely, since solutions are no longer assumed to be globally smooth functions, but are still smooth manifolds, one can associate the formation of shock waves with the formation of singularities of their projection to the space of independent variables [37]. This idea has also found an application in non-stationary filtration problems [38–40].

This beautiful geometric observation is accompanied with another concept of finding smooth and multivalued solutions of nonlinear PDEs, which is adding a differential constraint to the original system, such that the resulting overdetermined system turns out to be compatible [41] and its integration can be performed constructively by finding integrals of a completely integrable distribution. We discuss two ways of finding such constraints for the barotropic Euler system. One of them is based on specific geometry of Euler's equations in question, while the other is built up from the theory of differential invariants and quotient PDEs [42]. The first method is restricted to finding only constraints of the zeroth order involving independent and dependent variables, and the power of the second one is that it is universal in a sense that it is applicable to a much broader class of PDEs and allows one to find differential constraints of any order. However, we discuss both of the mentioned ways of finding constraints to show different aspects of the geometry that stands behind Euler's equations. This approach has found applications in integrability of many PDEs of mathematical physics, among them are two-dimensional gas dynamics [43], Khokhlov-Zabolotskaya equation [44], Hunter-Saxton equation [45], to name but a few.

The paper is organized as follows. We start with geometrical description of thermodynamic states of real gases using the formalism of contact and symplectic geometry and Lagrangian and Legendrian singularities. Then, we switch to Euler's equations in one spatial dimension and provide their geometrical description along with the structures arising from it. We discuss integrability of Euler's equations in terms of those structures finally finding classes of multivalued solutions, their caustics and shock wave fronts. We discuss solutions having one and two cusp singularities and their physical meaning. Further, we discuss differential invariants and get a quotient PDE for x-invariant differential constraints. We find certain thermodynamic state models for which we can get all x-invariant constraints. Using the same technique, we address barotropic Euler's equations in two and three dimensions, as well as fully compressible Euler's equations, and find explicit multivalued and smooth solutions in these cases.

2 Geometry and Thermodynamics

In this section, we give a geometrical description of thermodynamic properties of gases [46, 47] and show how theory of singularities can be used to model phase transitions. We consider only the case of gases, which is enough for the purposes of the paper, for more general studies see, for example, [48]. Here, the exposition mainly follows [13, 20].

2.1 Legendrian and Lagrangian Manifolds

Recall that any gas-like thermodynamic system in equilibrium is described by the two types of variables

- extensive variables: inner energy e, specific volume v, entropy s;
- intensive variables: temperature T, pressure p.

The energy conservation law states that expression

$$\theta = -ds + T^{-1}de + pT^{-1}dv = -ds + T^{-1}de - pT^{-1}\rho^{-2}d\rho, \tag{1}$$

where $\rho = v^{-1}$ is the density, must vanish. This statement that essentially comes from the physical world is contradictory from the mathematical viewpoint. To make this physical statement mathematically consistent, let us consider the contact manifold $\widehat{\Phi} = W \times W^* \times \mathbb{R} \simeq \mathbb{R}^5$, where $W \simeq \mathbb{R}^2(e, \rho)$, $W^* \simeq \mathbb{R}^2(p, T)$, and $s \in \mathbb{R}$ with structure form (1). Then, the energy conservation law implies that any thermodynamic system in equilibrium can be associated with a maximal integral submanifold $\widehat{L} \subset (\widehat{\Phi}, \theta)$ of the contact form θ, i.e.,

$$\theta|_{\widehat{L}} = 0.$$

The Legendrian submanifold $\widehat{L}$ is called *thermodynamic state*. In other words, the conservation of energy law holds on $\widehat{L}$.

In local coordinates (e, ρ) on $\widehat{L}$ we get

$$\widehat{L} = \left\{ s = S(e, \rho),\ T = \frac{1}{S_e},\ p = -\rho^2 \frac{S_\rho}{S_e} \right\} \tag{2}$$

for some function $S(e, \rho)$.

A natural question that arises at this point is how to build up a Legendrian submanifold $\widehat{L}$ having some experimental data on pressure, density, and temperature, since there are no devices that would help us with the measurement of entropy. To that end, let us consider projection

$$\pi : \widehat{\Phi} \to \Phi, \quad \pi(p, T, e, \rho, s) = (p, T, e, \rho).$$

Taking the differential of θ we equip the manifold Φ with the symplectic manifold structure

$$\Omega = d\theta = d(T^{-1}) \wedge de - d(pT^{-1}\rho^{-2}) \wedge d\rho.$$

Further, restriction of the map π to the Legendrian manifold $\widehat{L}$ gives a Lagrangian submanifold $L = \pi(\widehat{L}) \subset (\Phi, \Omega)$, such that

$$\Omega|_L = 0.$$

Any two-dimensional surface L in Φ can be given by two relations, called *state equations*

$$L = \{f(p, T, e, \rho) = 0,\ g(p, T, e, \rho) = 0\},$$

and since L is Lagrangian, functions f and g are not arbitrary but are in involution on L with respect to the Poisson bracket induced by the symplectic structure Ω:

$$\{f, g\}\ \Omega \wedge \Omega = df \wedge dg \wedge \Omega.$$

Condition $\{f, g\}|_L = 0$ is called *the compatibility condition for state equations*. Let (T, ρ) be local coordinates on L.

Theorem 3.1 *The Lagrangian manifold L is given by means of the Massieu-Planck potential $\phi(\rho, T)$*

$$p = -\rho^2 T\phi_\rho, \quad e = T^2\phi_T. \tag{3}$$

Proof If (T, ρ) are the local coordinates on L, then L is given as

$$L = \{p = P(T, \rho),\, e = E(T, \rho)\},$$

and applying the condition $\{f, g\}|_L = 0$, we get

$$(-\rho^{-2}T^{-1}P)_T = (T^{-2}E)_\rho,$$

and we finalize the proof by introducing the Massieu-Planck potential $\phi(t, \rho)$, as

$$-\rho^{-2}T^{-1}P = \phi_\rho, \quad T^{-2}E = \phi_T.$$

Now that we have a Lagrangian description of thermodynamic states, we can reconstruct L having the state equation $p = P(\rho, T)$. Moreover, using (2), one can find $S(e, \rho)$ and get a complete description of a thermodynamic system. Indeed, entropy is found to be

$$s = \phi + T\phi_T. \tag{4}$$

2.2 Riemannian Structures, Singularities, Phase Transitions

The symplectic manifold Φ is also equipped with the differential quadratic form

$$\kappa = d(T^{-1}) \cdot de - \rho^{-2}d(pT^{-1}) \cdot d\rho.$$

Its restriction to the Lagrangian surface L has to be negative, since $-\kappa|_L$ corresponds to the second central moment for a random vector of extensive variables. For more details we refer to [13, 20]. In other words, only those parts of the Lagrangian manifold L represent physical states, which are Riemannian with respect to $-\kappa|_L$. In the chosen coordinates (T, ρ) it takes the form

$$\kappa|_L = -(2T^{-1}\phi_T + \phi_{TT})dT \cdot dT + (2\rho^{-1}\phi_\rho + \phi_{\rho\rho})d\rho \cdot d\rho.$$

it is clearly seen from (3) that the condition $\kappa|_L < 0$ is

$$e_T > 0, \quad p_\rho > 0,$$

which is known to be the conditions of *thermodynamic stability* and reflects the naive expectation of physical systems behavior: the greater the density, the greater the pressure, and the greater the temperature, the greater the energy.

Here, we arrive at the crucial point of understanding phase transitions. Let us consider singularities of projections of the Lagrangian manifold L to the spaces of extensive variables $W \simeq \mathbb{R}^2(e, \rho)$ and intensive variables $W^* \simeq \mathbb{R}^2(p, T)$.

- projection to $W(e, \rho)$
  Singularities appear at points where

$$(de \wedge d\rho)|_L = 0 \Longleftrightarrow 2T^{-1}\phi_T + \phi_{TT} = 0 \Longleftrightarrow e_T = 0.$$

- projection to $W^*(p, T)$
  Singularities appear at points where

$$(dp \wedge dT)|_L = 0 \Longleftrightarrow 2\rho^{-1}\phi_\rho + \phi_{\rho\rho} = 0 \Longleftrightarrow p_\rho = 0.$$

Thus, singularities of projections of thermodynamic Lagrangian manifolds coincide with the set, where $\kappa|_L$ degenerates. Most of the models for real gases are constructed in such a way that (e, ρ) can serve as global coordinates on the corresponding Lagrangian manifold L. We therefore will not be interested in the appearance of singularities of the first type, leaving aside the discussion about their physical meaning. Having said that $p_\rho > 0$ allows us to sort out the points on L representing real physical states, one can expect that in the neighborhood of the points where this condition violates the behavior of a thermodynamic system is different from normal. Indeed, the system "jumps" into another point where $p_\rho > 0$, and this jump obeys certain laws.

Definition 3.1 A transition from a point $a_1 = (e_1, \rho_1) \in L$ to another point $a_2 = (e_2, \rho_2) \in L$ is called *phase transition*, if

- $p(a_1) = p(a_2)$
- $T(a_1) = T(a_2)$
- $\gamma(a_1) = \gamma(a_2)$,

where $\gamma = e - Ts + p/\rho$ is the Gibbs free energy.

Pairs of points (a_1, a_2) form a curve, which is called *coexistence curve*, or *phase transition curve*. Using the definition of a phase transition, we can write down its governing equations that directly follow from (3):

$$\begin{aligned} &p = -\rho_1^2 T\phi_\rho(T, \rho_1), \quad p = -\rho_2^2 T\phi_\rho(T, \rho_2), \\ &\phi(T, \rho_1) + \rho_1\phi_\rho(T, \rho_1) = \phi(T, \rho_2) + \rho_2\phi_\rho(T, \rho_2), \end{aligned} \tag{5}$$

2.3 Examples

Let us now consider examples of thermodynamic state models that will be used in the forthcoming sections.

Ideal gas

Ideal gas model is probably the first attempt to describe the properties of gas-like systems mathematically. It captures the main features of gas behavior that are observed far away from phase transition points. The Legendrian manifold is

$$\widehat{L} = \left\{ p = R\rho T,\ e = \frac{n}{2}RT,\ s = R\ln\left(\frac{T^{n/2}}{\rho}\right) \right\}, \tag{6}$$

where R is the universal gas constant, and n is the degree of freedom of a molecule.

The differential quadratic form $\kappa|_L$ is

$$\kappa|_{\widehat{L}} = -\frac{Rn}{2}\frac{dT^2}{T^2} - R\rho^{-2}d\rho^2.$$

Both inequalities $e_T > 0$, $p_\rho > 0$ are satisfied, and we do not observe phase transitions in this case.

van der Waals gas

The van der Waals model is probably the first that captures liquid-gas phase transitions and was constructed empirically by perturbation of an ideal gas state equations. Here, we use the reduced van der Waals state equations

$$\widehat{L} = \left\{ p = \frac{8T\rho}{3-\rho} - 3\rho^2,\ e = \frac{4nT}{3} - 3\rho,\ s = \ln\left(T^{4n/3}(3\rho^{-1} - 1)^{8/3}\right) \right\},$$

and the differential quadratic form $\kappa|_L$ is

$$\kappa|_{\widehat{L}} = -\frac{4n}{3T^2}dT^2 + \frac{6(\rho^3 - 6\rho^2 - 4T + 9\rho)}{\rho^2 T(\rho - 3)^2}d\rho^2.$$

It obviously changes its sign, and the Lagrangian manifold of van der Waals gas has a cusp-type singularity. It is shown in Fig. 1.

The phase transition curve divides the Legendrian manifold into three parts: gas phase, liquid phase, and wet steam. Points of the phase transition curve with the same values of pressure p and temperature T and different values of density $\rho_2 > \rho_1$ correspond to the liquid phase and the gas phase respectively, while points between ρ_1 and ρ_2 correspond to wet steam. Caustic and projection of the phase transition curve to $W^*(p, T)$ is shown in Fig. 2.

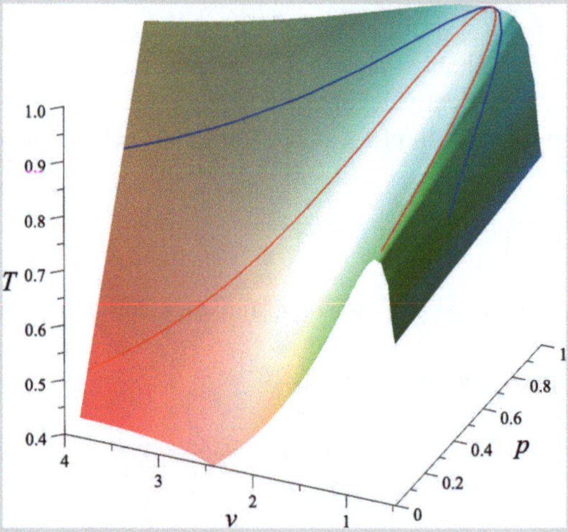

Fig. 1 Lagrangian manifold for van der Waals gas together with the set $p_\rho = 0$ (red), and coexistence curve (blue)

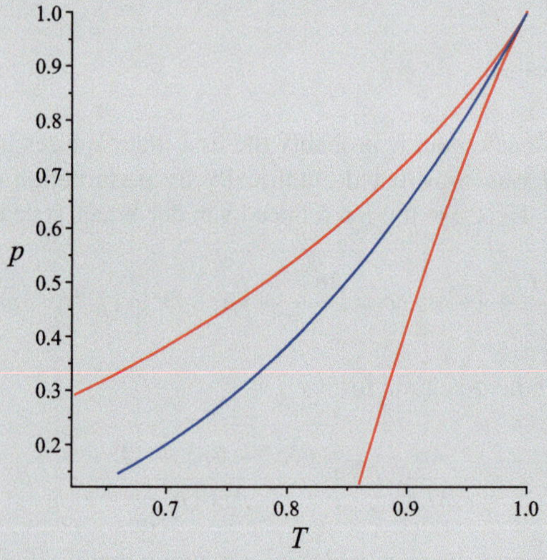

Fig. 2 Caustic (red) and phase transition curve (blue) for van der Waals gas

3 Barotropic 1D Euler's Equations

In this section, we discuss barotropic Euler's equations in one spatial dimension [1–3]. They can be seen as equations on either plane curves, or space curves in the absence of gravity [49].

We start with the fully compressible Euler's system in 1D, which consists of [4, 5]

- conservation of momentum

$$\rho(u_t + uu_x) = -p_x, \tag{1}$$

- conservation of mass

$$\rho_t + (\rho u)_x = 0, \tag{2}$$

- conservation of entropy along the flow

$$s_t + us_x = 0, \tag{3}$$

where $u(t, x)$ is the velocity, and x is a parameter on the curve.

Let the thermodynamic state of the fluid be given by some Legendrian manifold $\widehat{L}$. Fixing the entropy to be some constant $s(t, x) = s_0$, thus assuming the *isentropicity* of the flow, geometrically we get a curve on $l \subset \widehat{L}$ that can be parameterized by the density ρ. We can think of $l \subset L$ as of a *thermodynamic process*. Indeed, using (4) we have $s_0 = \phi + T\phi_T$, which can be seen as an equation for $T(\rho)$. By the implicit function theorem, there is a unique solution $T(\rho)$ due to the negativity of $\kappa|_L$. Plugging $T(\rho)$ into the state equations we also get $p(\rho, T(\rho)) = p(\rho)$. Under the assumptions made we get the two-component system, called *barotropic Euler's system*:

$$\mathcal{E} = \begin{cases} \rho_t + (\rho u)_x = 0, \\ u_t + uu_x + \dfrac{p'(\rho)}{\rho}\rho_x = 0. \end{cases} \tag{4}$$

3.1 Geometrical Structures Associated to Euler's Equations

Let $M = \mathbb{R}^2(t, x)$, $E = \mathbb{R}^4(t, x, u, \rho)$, and let $\Omega^2(E)$ be a module of differential 2-forms on E. We start with an observation that system (4) is a particular case of Jacobi-type systems [28] and can therefore be associated with two differential 2-forms on E in the following sense. Any differential 2-form $\omega \in \Omega^2(E)$ generates a differential operator

$$\Delta_\omega \colon C^\infty(M) \to \Omega^2(M), \quad \Delta_\omega(f) = \omega|_{\Gamma^0(f)},$$

where $\Gamma^0(f) \subset E$ is a graph of the vector-function f, $\omega|_{\Gamma^0(f)}$ is a restriction of the form ω to $\Gamma^0(f)$.

In case of system (4) the function $f = (u(t,x), \rho(t,x))$, and the forms are

$$\begin{cases} \omega_1 = \rho dt \wedge du + u dt \wedge d\rho - dx \wedge d\rho, \\ \omega_2 = u dt \wedge du + \dfrac{p'(\rho)}{\rho} dt \wedge d\rho - dx \wedge du. \end{cases} \tag{5}$$

System (4) can therefore be written as

$$\Delta_{\omega_1}(f) = 0, \quad \Delta_{\omega_2}(f) = 0.$$

Indeed, let $\Gamma^0(f) \subset E$ be locally given by

$$\Gamma^0(f) = \{u = u(t,x),\ \rho = \rho(t,x)\} \subset E.$$

Then, if we take, say, the differential form ω_1, we get

$$\Delta_{\omega_1}(f) = \omega_1|_{\Gamma^0(f)} = (\rho_t + (\rho u)_x)\, dt \wedge dx.$$

Note that in (5) the roles of dependent and independent variables are equalized.

Definition 3.2 A 2-dimensional manifold $N \subset E$ is said to be a *(multivalued) solution* of $\mathcal{E}$, if $\omega_1|_N = \omega_2|_N = 0$.

This provides a natural method of generalizing classical solutions to multivalued, staying in the smooth manifolds setting and reduces the problem of finding solutions for (4) to finding integral manifolds of differential forms (5), which are in general not globally given as graphs of functions and may have singularities of their projections to the space of independent variables. Also, one can notice that (5) are objects on the space of independent and dependent variables, not involving derivatives of unknown functions.

It is also worth mentioning that the form ω_2 is closed and non-degenerate, which implies that solutions to (4) are Lagrangian submanifolds $N \subset E$.

We observe that both phase transitions and shock waves have the same mathematical framework behind them. In the first case, it is singularities of projections of thermodynamic Lagrangian manifolds to the space of intensive variables, and in the second case it is singularities of projections of solutions thought of as Lagrangian manifolds to the space of independent variables.

Let us now show how one can use this geometric framework to find solutions constructively.

Having fixed the volume form on E as $q = dt \wedge dx \wedge du \wedge d\rho$ one can introduce a bilinear operator

$$P\colon \Omega^2(E) \times \Omega^2(E) \to C^\infty(E), \quad \alpha_1 \wedge \alpha_2 = P(\alpha_1, \alpha_2) q, \quad \alpha_i \in \Omega^2(E).$$

Let $P_\omega = \|P(\omega_i, \omega_j)\|$ be its matrix in the basis $\langle\omega_1, \omega_2\rangle$. Then, system $\mathcal{E}$ is said to be *hyperbolic*, if $\det(P_\omega) < 0$, *elliptic*, if $\det(P_\omega) > 0$ and *parabolic*, if $\det(P_\omega) = 0$. One can show straightforwardly that for (5)

$$P_\omega = \begin{pmatrix} 2\rho & 0 \\ 0 & -2\rho^{-1}p'(\rho) \end{pmatrix},$$

and we conclude that the condition for system (4) to be of hyperbolic type is $p'(\rho) > 0$, which is exactly the condition for negativity of $\kappa|_L$.

Theorem 3.2 *If a thermodynamic process curve $l \subset \widehat{L}$ lies in the domain of negativity of the form $\kappa|_L$, then the system $\mathcal{E}$ is hyperbolic.*

This underlines that negativity of the differential quadratic form $\kappa|_L$ is important both from thermodynamic perspective and for defining the type of the Euler equations.

Since system (4) is defined by the zeros of forms (5), the differential forms

$$\widehat{\omega_1} = a_{11}\omega_1 + a_{12}\omega_2,$$
$$\widehat{\omega_2} = a_{21}\omega_1 + a_{22}\omega_2,$$

where $a_{11}a_{22} - a_{12}a_{21} \neq 0$, are equally good for us. Moreover, in the hyperbolic case one can always choose *effective forms* ω_1 and ω_2, i.e., those satisfying [28]

$$\omega_1 \wedge \omega_2 = 0, \quad \omega_1 \wedge \omega_1 = -\omega_2 \wedge \omega_2. \tag{6}$$

It will be more convenient for us to work on the basis given by effective forms, and we have

Theorem 3.3 *Let $\mathcal{E}$ be a system of hyperbolic type. Then, it can be given by 2-forms*

$$\omega_1 = A(\rho)(\rho dt \wedge du + u dt \wedge d\rho - dx \wedge d\rho),$$
$$\omega_2 = u dt \wedge du + \rho A^2(\rho) dt \wedge d\rho - dx \wedge du,$$

where $A(\rho) = \rho^{-1}\sqrt{p'(\rho)}$, and 2-forms ω_1, ω_2 satisfy relations (6).

We will need also the operator A_ω acting in the module of vector fields $D(E)$ on E in the following way:

$$A_\omega\colon D(E) \to D(E), \quad X\rfloor\omega_2 = A_\omega(X)\rfloor\omega_1.$$

For the effective basis $\langle\omega_1, \omega_2\rangle$ the square of this operator is an identity, $A_\omega^2 = \mathrm{id}$, and its matrix in the basis $\langle\partial_t, \partial_x, \partial_\rho, \partial_u\rangle$ is

$$W = \frac{1}{\rho A(\rho)} \begin{pmatrix} u & -1 & 0 & 0 \\ u^2 - \rho^2 A^2(\rho) & -u & 0 & 0 \\ 0 & 0 & 0 & \rho A^2(\rho) \\ 0 & 0 & \rho & 0 \end{pmatrix}.$$

The operator $A_\omega(a)$, $a \in E$, has two real eigenspaces $C_+(a)$ and $C_-(a)$, and the tangent space $T_a E$ splits into a direct sum of eigenspaces $C_\pm(a)$:

$$T_a E = \mathcal{C}_-(a) \oplus \mathcal{C}_+(a).$$

Letting $a \in E$ running over E, we get *characteristic distributions* $C_\pm$ generated by $\langle X_-, Y_- \rangle$ and $\langle X_+, Y_+ \rangle$ correspondingly:

$$X_\pm = \pm A(\rho)\partial_u + \partial_\rho,$$
$$Y_\pm = (\mp \rho A(\rho) + u)^{-1}\partial_t + \partial_x,$$

which is a result of straightforward computations.

Theorem 3.4 *Distributions* $\mathcal{C}_+$ *and* $\mathcal{C}_-$ *are integrable if*

$$p(\rho) = c_0 \rho^3 + c_1, \tag{7}$$

where c_0 *and* c_1 *are constants.*

In the case when both characteristic distributions are integrable, one can perform the integration for the Cauchy problem explicitly [28]. However, the case (7) is well known and studied [5, 33–35], so we shall try to make a progress for an unspecified function $p(\rho)$. To that end, we will need the following theorem [28].

Theorem 3.5 *A two-dimensional manifold* $N \subset E$ *is a multivalued solution to* $\mathcal{E}$, *if and only if the tangent spaces* $T_a N$ *for all* $a \in N$ *have one-dimensional intersections* $h_+(a)$ *and* $h_-(a)$ *with planes* $\mathcal{C}_+(a)$ *and* $\mathcal{C}_-(a)$*:*

$$T_a N = h_+(a) \oplus h_-(a).$$

The lines $h_\pm(a) = \mathcal{C}_\pm(a) \cap T_a N$ are called *characteristic directions*, see Fig. 3. Using theorem 3.5, we can find solutions in the following way. Let

$$N_0 = \{F(t, x, u, \rho) = 0\} \subset E,$$

where the function $F(t, x, u, \rho)$ (we will call it *constraint*) is to be found and take two vector fields $V_\pm \in \mathcal{C}_\pm$, that are

- tangent to N_0, i.e., $V_\pm \rfloor dF = 0$
- their restrictions $h_+ = V_+|_{N_0}$ and $h_- = V_-|_{N_0}$ define an integrable distribution on N_0, i.e.,

$$V_+ \wedge V_+ \wedge [V_+, V_-] = 0. \tag{8}$$

Fig. 3 Characteristic directions

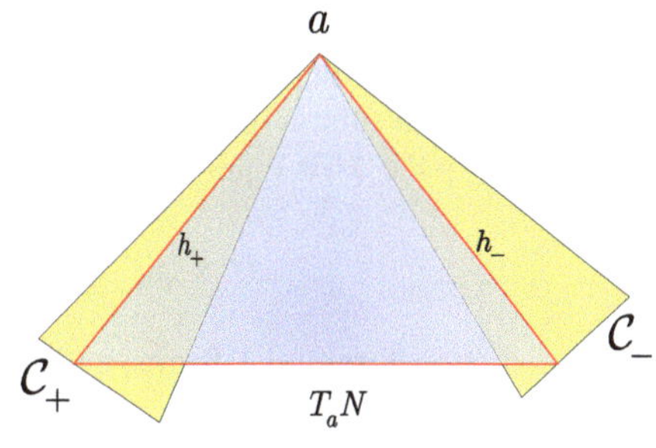

Then, finding F from (8), we get a two-dimensional integrable distribution $h = \langle h_+, h_- \rangle$, and according to Theorem 3.5, it spans a multivalued solution N.

Theorem 3.6 *Let function $F(t, x, u, \rho)$ be x-independent, i.e.,*

$$F(t, x, \rho, u) = t - f(u, \rho)/\rho.$$

Then, the distribution $V = \langle V_+, V_- \rangle$ is integrable, if the function $f(u, \rho)$ satisfies the PDE

$$f_{uu} - A^{-2}(\rho) f_{\rho\rho} = 0. \tag{9}$$

This theorem is a result of straightforward computations. Equation (9) is a linear wave equation with non-constant "wave speed". It can be reduced to the wave equation with constant coefficient for some particular barotropic functions $p(\rho)$.

Theorem 3.7 *Equation (9) is equivalent to the wave equation with constant coefficients if*

$$A(\rho) = (\beta_1 \rho + \beta_2)^{-2},$$

where β_1, β_2 are constants.

Since we know the structure of solutions to linear wave equations, we can give a complete description of x-invariant constraints in this case.

3.2 Integrability via Quotients

We observe that the main trick that allows us to find solutions explicitly is to add an additional relation (constraint) to the original system, in a way that makes the distribution spanning the multivalued solution integrable. Let us try to look at this scheme from a more general perspective. Namely, we will show that (9) is a quotient PDE for (4) with respect to x-translations. Exposition here is informal, given in coordinates, and is limited to the purposes of this paper, for more comprehensive discussion and a coordinate-free way of definition of all necessary constructions we refer to [42].

Let $x = (x_1, \dots, x_n)$ be the collection of independent variables, and $u = (u^1, \dots, u^m)$ be the collection of dependent variables. Recall that any system of PDEs or order k can be seen as a submanifold $\mathcal{E} \subset J^k(\mathbb{R}^n) = J^k$ in the space of k-jets of functions on $\mathbb{R}^n$. The space J^k is equipped with the Cartan distribution $\mathcal{C}$

$$\mathcal{C}\colon J^k \ni a \mapsto \mathcal{C}_a \subset T_a J^k,$$

generated by differential 1-forms, called *Cartan forms*, that in canonical jet coordinates (x, u^j_σ) are written as

$$\omega^j_\sigma = du^j_\sigma - \sum_{i=1}^n u^j_{\sigma i} dx_i, \quad |\sigma| \leq k-1.$$

The restriction $\mathcal{C}_\mathcal{E}$ of the Cartan distribution $\mathcal{C}$ to the submanifold $\mathcal{E}$ is

$$\mathcal{C}_\mathcal{E}\colon \mathcal{E} \ni a \mapsto \mathcal{C}_\mathcal{E}(a) = T_a\mathcal{E} \cap \mathcal{C}_a.$$

By *(generalized or multivalued) solutions* we mean n-dimensional integral manifolds of the restriction $\mathcal{C}_\mathcal{E}$ of the Cartan distribution to $\mathcal{E}$.

A vector field $X \in D(\mathcal{E})$ is said to be an *infinitesimal symmetry* of $\mathcal{E}$, if it preserves the distribution $\mathcal{C}_\mathcal{E}$, i.e., $[X, \mathcal{C}_\mathcal{E}] \subset \mathcal{C}_\mathcal{E}$. All such vector fields form a Lie algebra $\mathfrak{g}$, called the symmetry Lie algebra.

Let $\mathcal{E} \subset J^k$, let $\mathcal{E}_r \subset J^r$, $r \geq k$ be its prolongation.

Definition 3.3 A function $I \in C^\infty(\mathcal{E}_r)$ is said to be a *differential* $\mathfrak{g}$*-invariant of order* r, if it is constant on $\mathfrak{g}$-orbits, i.e.,

$$X^r(I) = 0$$

for all $X \in \mathfrak{g}$.

Functions $a_1, \dots, a_n \in C^\infty\left(J^l\right)$ are said to be *in general position*, if

$$\widehat{d}a_1 \wedge \dots \wedge \widehat{d}a_n \neq 0$$

in some Zariski-open set in J^{l+1}, where $\widehat{d}$ is the total differential:

$$\widehat{d}f = \sum_{i=1}^n \frac{df}{dx_i} dx_i.$$

Let us choose $\mathfrak{g}$-invariants $a_1, \dots, a_n$ in general position and construct *Tresse derivatives* as follows:

$$\frac{d}{da_i} = \sum_j A_{ij} \frac{d}{dx_j}, \quad \frac{da_j}{da_i} = \delta_{ij},$$

where

$$\frac{d}{dx_i} = \frac{\partial}{\partial x_i} + \sum_{j,\sigma} u^j_{\sigma i} \frac{\partial}{\partial u^j_\sigma},$$

and δ_{ij} is the Kronecker delta. They are partial derivatives with respect to invariants a_i.

Theorem 3.8 *Let $a_1, \ldots, a_n$ be $\mathfrak{g}$-invariants of order $\leq r$ in general position. Then, for any $\mathfrak{g}$-invariant a of order $\leq r$ the Tresse derivatives $\frac{da}{da_i}$ are $\mathfrak{g}$-invariants of order $\leq r + 1$.*

The following important result known as the *global Lie-Tresse theorem* [42] gives conditions for the field of rational $\mathfrak{g}$-invariants to be finitely generated.

Theorem 3.9 *Let $\mathcal{E}_k \subset J^k$ be an algebraic formally integrable differential equation and let $\mathfrak{g}$ be its algebraic symmetry Lie algebra. Then, there exist rational differential $\mathfrak{g}$-invariants $a_1, \ldots, a_n, b^1, \ldots, b^N$ of order $\leq l$, such that the field of rational $\mathfrak{g}$-invariants is generated by rational functions of these functions and Tresse derivatives $\frac{d^{|\alpha|} b^j}{da^\alpha}$.*

Quotient PDEs can be identified with relations between differential invariants, called *syzygies*, in the following sense. Let the algebra of differential invariants be generated by $a_1, \ldots, a_n, b^1, \ldots, b^N$. Their restriction to some solution s of $\mathcal{E}$ gives $(n + N)$ functions on $\mathbb{R}^n$, which means that there are at least N relations between them locally expressed as

$$b^j|_s = B^j(a_1|_s, \ldots, a_n|_s), \quad j = 1, \ldots, N.$$

Since $a_1, \ldots, a_n, b^1, \ldots, b^N$ are $\mathfrak{g}$-invariants, functions B^j are determined by the $\mathfrak{g}$-orbit of s, rather than by s itself.

We shall find functions $B^j(a_1|_s, \ldots, a_n|_s)$ as solutions to a certain PDE. To that end, let us differentiate $b^j = B^j(a_1, \ldots, a_n)$ in the sense of Tresse to get

$$J_{ji} = \frac{db^j}{da_i} = B^j_{a_i}(a_1, \ldots, a_n), \ldots, J_{j\alpha\beta} = \frac{d^2 b^j}{da_\alpha da_\beta} = B^j_{a_\alpha a_\beta}(a_1, \ldots, a_n), \ldots$$

By Theorem 3.8, functions $J_{ji}, J_{j\alpha\beta}, \ldots$ are $\mathfrak{g}$-invariants. Finding syzygies between them, one gets

$$F(a_1, \ldots, a_n, B^1, \ldots, B^N, B^j_{a_i}, B^j_{a_\alpha a_\beta}, \ldots) = 0, \tag{10}$$

for some (probably, a number of) function F. Equation (10) is called a *quotient PDE*. The two most important features of quotient PDEs are

- Let $b^j = B^j(a)$ be a solution to a quotient PDE. Then, adding differential constraints $b^j - B^j(a) = 0$ gives us a finite type equation $\widehat{\mathcal{E}} = \mathcal{E} \cap \{b^j - B^j(a) = 0\}$ with solutions being a $\mathfrak{g}$-orbit of a solution to $\mathcal{E}$.

- Overdetermined system $\widehat{\mathcal{E}} = \mathcal{E} \cap \{b^j - B^j(a) = 0\}$ has the same symmetry Lie algebra as the original $\mathcal{E}$ has.

Thus, adding differential constraints found by means of quotient PDEs, one reduces the integration of the PDE to finding integrals of a completely integrable Cartan distribution with the same symmetry Lie algebra.

After this brief description of the concept of quotient PDEs we will show how these ideas can be applied to the Euler system (4).

Theorem 3.10 *The symmetry Lie algebra of the system of the Euler equations* (4) *is generated by vector fields*

$$X_1 = \frac{\partial}{\partial x}, \quad X_2 = \frac{\partial}{\partial t}, \quad X_3 = t\frac{\partial}{\partial x} + \frac{\partial}{\partial u}, \quad X_4 = t\frac{\partial}{\partial t} + x\frac{\partial}{\partial x},$$
$$X_5 = C_1 e^{-C_1 u} Y \frac{\partial}{\partial t} + (C_1 Y u + Y'\rho + Y) e^{-C_1 u} \frac{\partial}{\partial x},$$

where $Y = Y(\rho)$ is a solution to the ODE

$$-\rho A^2(\rho) C_1^2 Y + \rho Y'' + 2Y' = 0,$$

and C_1 is a constant.

Let $\mathfrak{g} = \langle \partial_x \rangle$ and let

$$a_1 = u, \quad a_2 = \rho, \quad b^1 = t$$

be the generators of the algebra of $\mathfrak{g}$-invariants. The Tresse derivatives are

$$\frac{d}{du} = \frac{\rho_x}{\rho u_x^2 - \rho A^2 \rho_x^2} \frac{d}{dt} + \frac{\rho u_x + u\rho_x}{\rho u_x^2 - \rho A^2 \rho_x^2} \frac{d}{dx},$$
$$\frac{d}{d\rho} = \frac{u_x}{\rho A^2 \rho_x^2 - \rho u_x^2} \frac{d}{dt} + \frac{A^2 \rho\rho_x + u u_x}{\rho A^2 \rho_x^2 - \rho u_x^2} \frac{d}{dx}.$$

Applying the Tresse derivatives to $t = B(u, \rho)$ we get

$$\frac{dB}{du} = \frac{\rho_x}{u_x^2 \rho - A^2 \rho\rho_x^2}, \quad \frac{dB}{d\rho} = \frac{u_x}{A^2 \rho\rho_x^2 - u_x^2 \rho}, \tag{11}$$

$$\frac{d^2 B}{du^2} = \frac{1}{(u_x^2 \rho - \rho_x^2 \rho A^2)^3} \big(2(A^2 + 2\rho A A')\rho\rho_x^4 u_x + \\ + (-\rho\rho_x^3 u_{xx} + 3\rho\rho_x^2 \rho_{xx} u_x - 2\rho_x^4 u_x)\rho A^2 + u_x^2 \rho^2 (-3\rho_x u_{xx} + \rho_{xx} u_x)\big), \tag{12}$$

$$\frac{d^2 B}{d\rho^2} = \frac{1}{(-u_x^2 \rho + \rho_x^2 \rho A^2)^3} \big(-2\rho A^2 (A^2 + 2\rho A A')\rho_x^4 u_x + \rho^2 A^4 \rho_x^3 u_{xx} - \\ - 3\rho^2 A^4 \rho_x^2 \rho_{xx} u_x + 3\rho^2 A^2 \rho_x u_x^2 u_{xx} - \rho^2 A^2 \rho_{xx} u_x^3 + 4\rho A^2 \rho_x^2 u_x^3 - 2\rho u_x^5\big) \tag{13}$$

$$
\begin{aligned}
\frac{d^2B}{dud\rho} = \frac{1}{(u_x^2\rho - \rho_x^2\rho A^2)^3}\big(&(-u_x^2\rho\rho_x^3 - \rho_x^5\rho A^2)(A^2 + 2\rho A(\rho)A') - \\
&-\rho^2A^4\rho_x^3\rho_{xx} + (3\rho\rho_x^2u_xu_{xx} - 3\rho\rho_x\rho_{xx}u_x^2 + 3\rho_x^3u_x^2)\rho A^2 - \\
&-\rho\rho_xu_x^4 + \rho^2u_x^3u_{xx}\big).
\end{aligned} \tag{14}
$$

Finding the syzygy between (11)–(14), we get a quotient PDE

$$
A^2\rho\frac{d^2B}{du^2} - \rho\frac{d^2B}{d\rho^2} - 2\frac{dB}{d\rho} = 0,
$$

which for $B(u, \rho) = f(u, \rho)/\rho$ becomes

$$
f_{uu} - A^{-2}f_{\rho\rho} - 0, \tag{15}
$$

and we thus have rederived (9) as a quotient PDE for (4) with respect to $\mathfrak{g} = \langle\partial_x\rangle$.

Let

$$
J^1 \supset \widehat{\mathcal{E}} = \mathcal{E} \cap \left\{t - f(u, \rho) = 0,\ \frac{d}{dt}(t - f(u, \rho)) = 0,\ \frac{d}{dx}(t - f(u, \rho)) = 0\right\},
$$

where $f(u, \rho)$ is a solution to (15). Then, $\dim\widehat{\mathcal{E}} = 3$, and therefore $\operatorname{codim}\mathcal{C}_{\widehat{\mathcal{E}}} = 1$.

Let (x, u, ρ) be the local coordinates on $\widehat{\mathcal{E}}$. Then, the Cartan distribution $\mathcal{C}_{\widehat{\mathcal{E}}}$ is generated by

$$
\begin{aligned}
\omega = &-\frac{\rho(f_\rho\rho - f)}{\rho^2A^2f_u^2 - f_\rho^2\rho^2 + 2f_\rho f\rho - f^2}dx + \frac{(f_\rho\rho - f)(f_uu - f_\rho\rho + f)}{\rho^2A^2f_u^2 - f_\rho^2\rho^2 + 2f_\rho f\rho - f^2}du + \\
&+\frac{(-\rho^2A^2f_u + f_\rho\rho u - fu)(f_\rho\rho - f)}{\rho^3A^2f_u^2 - \rho^3f_\rho^2 + 2\rho^2ff_\rho - f^2\rho}d\rho,
\end{aligned} \tag{16}
$$

and it is easy to check that $\mathcal{C}_{\widehat{\mathcal{E}}}$ is integrable for any solution $f(u, \rho)$ to (15). Moreover, ∂_x is a symmetry of (16), and therefore by the Lie-Bianchi theorem [24, 28] integrals for $\widehat{\mathcal{E}}$ are given by

$$
t = \frac{f(u, \rho)}{\rho},\quad \int\frac{\omega}{\omega(\partial_x)} = \text{const}.
$$

Remark 3.1 The distribution generated by ω coincides with that generated by $\langle h_-, h_+\rangle$.

3.3 Separation Ansatz $f(u, \rho) = \rho\mu(u) + \nu(\rho)$

Let us now elaborate on the quotient PDE (15). Looking for the solution in the form $f(u, \rho) = \rho\mu(u) + \nu(\rho)$, one gets

$$\mu'' = \lambda, \quad \nu'' = \lambda\rho A^2,$$

for some constant λ, and we get the first integral for $\mathcal{E}$:

$$t = \frac{\lambda u^2}{2} + \alpha_0 u + \frac{\alpha_2}{\rho} + \frac{\lambda}{\rho}\int Q(\rho)d\rho + t_0, \tag{17}$$

where α_j, t_0 are constants, $Q(\rho) = \int \rho A^2 d\rho$.

To find the second integral, one has to integrate the form

$$\frac{\omega}{\omega(\partial_x)} = dx - \left(\lambda u^2 + \alpha_0 u - \lambda Q(\rho) + \frac{\lambda}{\rho}\int Q(\rho)d\rho + \frac{\alpha_2}{\rho}\right)du + \\ + \left(\frac{\lambda u}{\rho^2}\int Q(\rho)d\rho - \frac{\lambda u}{\rho}Q(\rho) + \rho A^2(\rho)(\lambda u + \alpha_0) + \frac{\alpha_2 u}{\rho^2}\right)d\rho,$$

and the second integral is

$$x = \frac{\lambda u^3}{3} - \lambda u Q(\rho) + \frac{\alpha_0 u^2}{2} + \frac{\lambda u}{\rho}\int Q(\rho)d\rho + \frac{\alpha_2 u}{\rho} - \alpha_0 Q(\rho) + x_0, \tag{18}$$

where x_0 is a constant.

Formulas (17)–(18) give a solution N of $\widehat{\mathcal{E}}$ for any thermodynamic state function $p(\rho)$. One can see that independent and dependent variables switch their roles, which reflects the idea that solutions are manifolds rather than functions.

Singularities of projection of N to M occur at the points where $(dx \wedge dt)|_N = 0$ forming a curve called *caustic*. Caustic for (17)–(18) is

$$x(\rho) = x_0 - \frac{Z_\pm^3}{3\lambda^2 A^3\rho^6} + \frac{\alpha_0 Z_\pm^2}{2\lambda^2 A^2\rho^4} \pm \frac{\alpha_0 Z_\pm}{\lambda\rho} \mp \frac{Z_\pm^2}{\lambda A\rho^3} - \alpha_0 Q,$$
$$t(\rho) = \frac{Z_+^2}{2\lambda A^2\rho^4} + \frac{Z_+}{\rho}\left(1 - \frac{\alpha_0}{\lambda\rho A}\right) - \alpha_0\rho A + \lambda Q + t_0,$$

where

$$Z_\pm = \alpha_0\rho^2 A \mp \lambda\rho Q \pm \lambda\int Q d\rho \pm \alpha_2.$$

From the physical perspective, a multivalued solution makes little sense, since it is not possible to observe different values of the density at the same point (t, x). The way out here is to cut a multivalued solution transforming it into a single-valued but discontinuous function. This cutting is performed in a way similar to the one we have used to find a phase transition curve for thermodynamic Lagrangian manifolds previously. Namely, the points of discontinuity are found from the following conditions:

- $x(u_1, \rho_1) = x(u_1, \rho_1)$,
- $t(u_1, \rho_1) = t(u_1, \rho_1)$,
- $H(u_1, \rho_1) = H(u_1, \rho_1)$,

where $H(u, \rho)$ is some conservation law, which in our case will be the mass conservation law following [5]. We see the obvious analogy to the phase transitions picture in thermodynamics: independent variables (x, t) play the role of intensive variables (p, T), and unknown functions (u, ρ) play the role of extensive variables (e, ρ). The mass conservation law $H(u, \rho)$ is a counterpart of the Gibbs potential γ, which is dual to the mass variable [13]. The discontinuity curve found from the above conditions is called *shock wave front*.

Let now (t, ρ) be coordinates on N. Resolving (17) with respect to u, we get

$$u = U_{\pm}(\rho, t) = -\frac{\alpha_0}{\lambda} \pm \frac{1}{\lambda\rho}\sqrt{2\rho\left(\rho\lambda(t - t_0) + \frac{\rho\alpha_0^2}{2} - \lambda\alpha_2 - \lambda^2 \int Qd\rho\right)}. \quad (19)$$

The mass conservation law is a 1-form Θ which is closed on any solution N. We therefore get

$$\Theta_{\pm} = \rho dx - \rho U_{\pm} dt.$$

Locally $\Theta_{\pm}|_N = dH_{\pm}$, and the function $H_{\pm}(\rho, t)$ is

$$H_{\pm}(\rho, t) = \mp\frac{1}{\lambda\sqrt{\rho}}\sqrt{2\left(\rho\lambda(t - t_0) + \frac{\rho\alpha_0^2}{2} - \lambda\alpha_2 - \lambda^2 \int Qd\rho\right)}\left(\rho\lambda Q - \lambda \int Qd\rho - \alpha_2\right).$$

3.3.1 Solutions for an Ideal Gas

Let us now illustrate how solutions obtained previously for an arbitrary function $p(\rho)$ look like in the case of an ideal gas. Using (6), we get

$$T(\rho) = \exp\left(\frac{2s_0}{Rn}\right)\rho^{2/n}, \quad p(\rho) = R\exp\left(\frac{2s_0}{Rn}\right)\rho^{2/n+1},$$

and the function $A(\rho) = \rho^{-1}\sqrt{p'(\rho)} = A_0\rho^m$ (see also [1, 2]), where

$$A_0 = \sqrt{R\left(1 + \frac{2}{n}\right)\exp\left(\frac{2s_0}{Rn}\right)}, \quad m = \frac{1}{n} - 1.$$

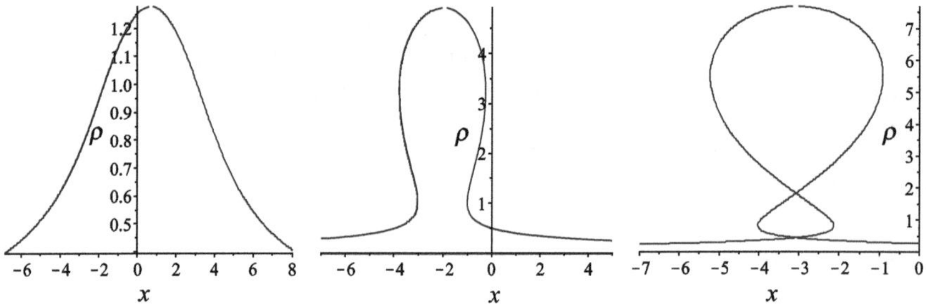

Fig. 4 Density graph for $n = 3$ at moments $t = 0$, $t = 2.7$, $t = 3.75$

The integrals (17)–(18) are

$$t = \frac{\lambda u^2}{2} + \alpha_0 u + \frac{\alpha_2}{\rho} + \frac{\lambda A_0^2 \rho^{2m+2}}{2(2m^2 + 5m + 3)} + t_0,$$

$$x = \frac{1}{\rho(2m^2 + 5m + 3)} \left((m+1)(2m+3) \left(\alpha_2 u + \rho \left(\frac{\lambda u^3}{3} + \frac{\alpha_0 u^2}{2} + x_0 \right) \right) - \right.$$
$$\left. - A_0^2 \rho^{2m+3} \left(m(\lambda u + \alpha_0) + \lambda u + \frac{3\alpha_0}{2} \right) \right).$$

Parametric equations for the caustics are

$$x(\rho) = x_0 - \frac{Z_\pm^3}{3\lambda^2 A_0^3 \rho^{3m+6}} + \frac{\alpha_0 Z_\pm^2}{2\lambda^2 A_0^2 \rho^{2m+4}} \pm \frac{\alpha_0 Z_\pm}{\lambda \rho} \mp \frac{Z_\pm^2}{\lambda A_0 \rho^{m+3}} - \frac{\alpha_0 A_0^2 \rho^{2m+2}}{2(m+1)},$$
$$t(\rho) = \frac{Z_+^2}{2\lambda A_0^2 \rho^{2m+4}} + \frac{Z_+}{\rho} \left(1 - \frac{\alpha_0}{\lambda A_0 \rho^{m+1}} \right) - \alpha_0 A_0 \rho^{m+1} + \frac{\lambda A_0^2 \rho^{2m+2}}{2(m+1)} + t_0,$$

where

$$Z_\pm = \alpha_0 A_0 \rho^{m+2} \mp \frac{\lambda A_0^2 \rho^{2m+3}}{2m+3} \pm \alpha_2.$$

The graphs of the multivalued solutions are presented in Fig. 4, where the constants are put $A_0 = 1,\ \alpha_0 = 1,\ t_0 = 1,\ \alpha_2 = -2,\ \lambda = 1,\ x_0 = 0$.

One can observe that evolution of initially smooth solution leads to the formation of two cusps that move toward one another. The shock wave fronts for these two cusps are depicted in Fig. 5.

3.3.2 Solutions for Van Der Waals Gas

Here, we elaborate solution of the van der Waals case, where we also have phase transitions along with shock waves. First, having a phase transition curve in the space

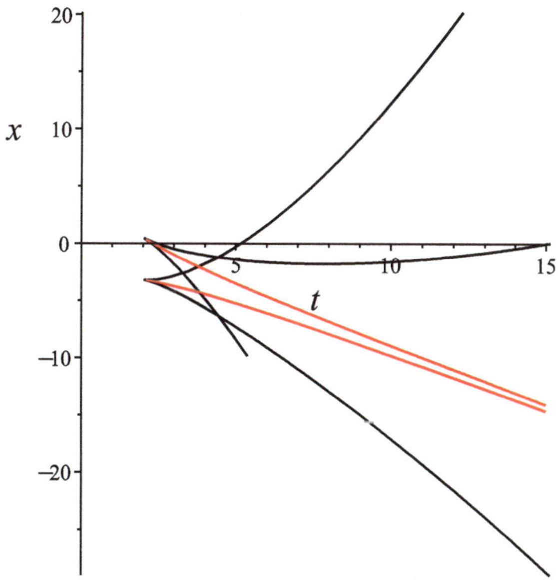

Fig. 5 Caustics (black) and shock wave fronts (red)

of thermodynamic variables we can reproduce it in the space of independent variables, because having a solution we get all the thermodynamic quantities as functions of t and x, and thus we are able to identify the areas in $M(t, x)$ for different phases of the medium. The pressure function for van der Waals gases is

$$p(\rho) = 8 \exp\left(\frac{3s_0}{4n}\right)\left(\frac{3}{\rho} - 1\right)^{-1-2/n} - 3\rho^2. \tag{20}$$

For simplicity, we will concentrate here on solution with one cusp singularity in contrast to the ideal gas case where we had two cusps. To that end, we put $\lambda = 0$, and also keep in mind that if $f(u, \rho)$ is a solution to (15), then $f(u, \rho) + \alpha u + \beta$ is so as well for some constants α, β. Thus the first quadrature is [3]

$$t + \alpha_2 + \frac{C_2 - u}{\alpha_1 \rho} + \frac{C_3 u}{\alpha_1} = 0, \tag{21}$$

and the second one is

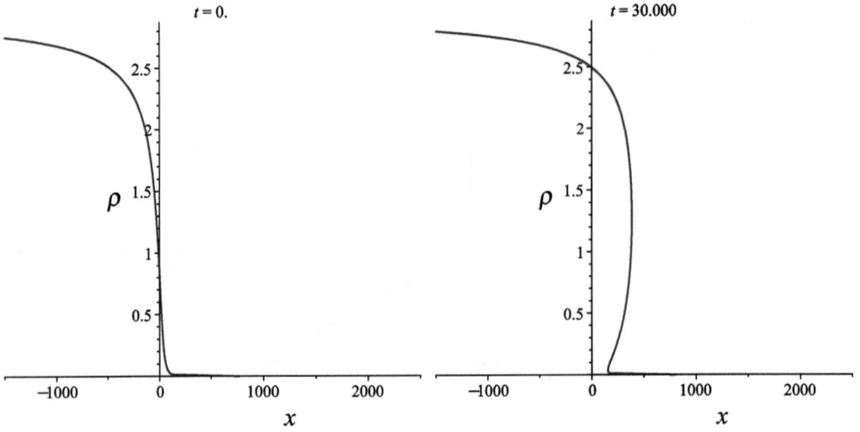

Fig. 6 Density graph for $n = 3$ at time moments $t = 0$, $t = 30$

$$
\begin{aligned}
0 = x + \alpha_3 + \frac{1}{\alpha_1}\Bigg(& C_1 \ln\rho - C_1 C_3 \rho + \frac{C_3 u^2}{2} + \frac{u(C_2 - u)}{\rho} - \\
& - C_5 \left(C_3 + \frac{C_7}{\rho}\right)^{C_6+1} \frac{2\rho^2 C_3^2 - C_7^2 (C_6+1)(C_3\rho(C_6+3) - C_6 - 2)}{(C_6+1)(C_6+2)(C_6+3) C_7^3 \rho^2} - \\
& - C_5 \left(C_3 + \frac{C_7}{\rho}\right)^{C_6+1} \frac{C_3 C_7 \rho (C_3 \rho (C_6+3) - 2C_6 - 2)}{(C_6+1)(C_6+2)(C_6+3) C_7^3 \rho^2}\Bigg),
\end{aligned} \tag{22}
$$

where the constants are

$$
\begin{aligned}
C_1 = -6, \quad C_3 = -1, \quad C_5 = 24\left(1 + \frac{2}{n}\right)\exp\left(\frac{3s_0}{4n}\right), \\
C_6 = -2 - \frac{2}{n}, \quad C_7 = 3.
\end{aligned} \tag{23}
$$

The graph of the multivalued solution is shown in Fig. 6.

Equations for the caustic are

$$
t(\rho) = -\alpha_2 - \frac{C_2 C_3}{\alpha_1} \pm \frac{(C_3\rho - 1)^2}{\alpha_1 \rho^2} \sqrt{C_1 \rho^3 + C_5 \left(C_3 + \frac{C_7}{\rho}\right)^{C_6}}, \tag{24}
$$

$$
\begin{aligned}
x(\rho) = & -\frac{1}{2\alpha_1}\left(2C_1 \ln \rho + C_1(C_3^3\rho^3 - 4\rho^2C_3^2 + 3C_3\rho - 2) + C_3C_2^3 + 2\alpha_1\alpha_3\right) \pm \\
& \pm \frac{C_2(C_3\rho - 1)^2}{\alpha_1\rho^2}\sqrt{C_1\rho^3 + C_5\left(C_3 + \frac{C_7}{\rho}\right)^{C_6}} - \\
& - \frac{C_5\left(C_3 + \frac{C_7}{\rho}\right)^{C_6}}{2(C_6+2)(C_6+3)C_7^3\alpha_1(C_6+1)\rho^3} \cdot \\
& \cdot \left(C_3^3(-4 + C_7^3(C_6^3 + 6C_6^2 + 11C_6 + 6) + (-2C_6 - 6)C_7)\rho^3 - \right. \\
& - 2C_7((2(C_6^3 + 6C_6^2 + 11C_6 + 6))C_7^2 + (-C_6^2 - 3C_6)C_7 - C_6)C_3^2\rho^2 + \\
& + C_7^2(C_6+1)((C_6+3)(5C_6+12)C_7 - 2C_6)C_3\rho - \\
& \left. - 2C_7^3(C_6+4)(C_6+2)(C_6+1)\right).
\end{aligned}
$$

Finally, the potential $H(\rho, t)$ is

$$
\begin{aligned}
H(\rho, t) = & \frac{\rho}{2\alpha_1(C_3\rho - 1)^2}\left(C_1C_3^3\rho^3 - 4C_1C_3^2\rho^2 + \right. \\
& \left. + \rho\left(C_2^2C_3^2 + (2C_2(t + \alpha_2)\alpha_1 + 5C_1)C_3 + \alpha_1^2(t + \alpha_2)^2\right) - 2C_1\right) - \\
& - \frac{C_5\left(C_3 + \frac{C_7}{\rho}\right)^{C_6}}{(C_6+2)\alpha_1C_7^2(C_6+1)\rho^2}(C_3\rho + C_7) \cdot \\
& \cdot (C_3(1 + (C_6+2)C_7)\rho - (C_6+1)C_7).
\end{aligned}
$$

Caustics and a shock wave front are shown in Fig. 7 Shock wave front together with phase transition curve are shown in Fig. 8. One can observe that we have two different phases on different sides of the discontinuity line, pure gas and wet steam.

3.4 Separation Ansatz $f(u, \rho) = g(\rho)h(u)$

Applying the standard separation ansatz $f(u, \rho) = \nu(\rho)\mu(u)$ to (15) we get

$$
\mu'' = \lambda\mu, \quad \nu'' + w(\rho)\nu = 0,
$$

where $w(\rho) = -\lambda A^2(\rho)$.

The function $\nu(\rho)$ satisfies the Schrödinger type equation, where the pressure function $p(\rho)$ is related to the potential $w(\rho)$ via $w(\rho) = -\lambda\rho^{-2}p'(\rho)$. Thus investigating the integrability Schrödinger type equations depending on the potential $w(\rho)$ we can find out integrable cases for (15), and consequently for (4).

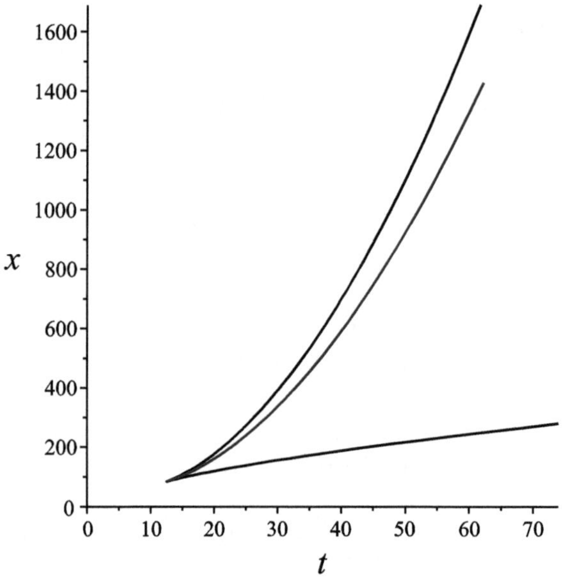

Fig. 7 Caustic (black) and shock wave front (red), $n = 3$

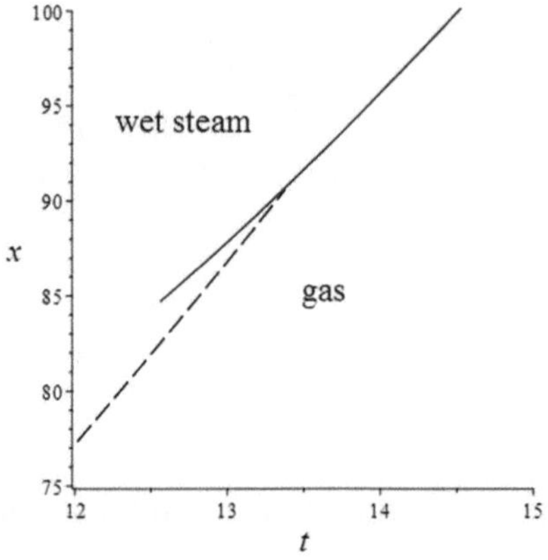

Fig. 8 Phase transition curve (dash line) and shock wave front (red line)

3.4.1 Integrability of Schrödinger Type Equations

Here we will show how to find integrable potentials for Schrödinger type equations, i.e., those for which one can write solution explicitly following [50, 51].

First, we observe that an ODE of the form

$$\mathcal{E} = \{u_k = f(x, u_0, \ldots, u_{k-1})\} \subset J^k$$

defines a smooth manifold $\mathcal{E}$ with the 1-dimensional Cartan distribution $\mathcal{C}_{\mathcal{E}}$ generated by

$$\omega_j = du_j - u_{j+1}dx, \quad j = 0, \ldots, k-2,$$
$$\omega_{k-1} = du_{k-1} - f dx,$$

or, equivalently, by the vector field

$$X_f = \mathcal{D} + \mathcal{D}(f)\partial_{u_k},$$

where

$$\mathcal{D} = \partial_x + u_1\partial_{u_0} + \cdots + u_k\partial_{u_{k-1}}.$$

Since an infinitesimal symmetry X satisfies $[X, \mathcal{C}_{\mathcal{E}}] \subset \mathcal{C}_{\mathcal{E}}$, it is clear that all $X \in \mathcal{C}_{\mathcal{E}}$ are infinitesimal symmetries, called *characteristic symmetries*. They form an ideal $\mathfrak{I} \subset \mathfrak{g}$, and the elements of the quotient $\mathfrak{g}/\mathfrak{I}$ are called *shuffle symmetries* [24, 25, 28].

Theorem 3.11 *Shuffle symmetries of the ODE $\mathcal{E}$ are of the form*

$$X_\phi = \sum_{j=0}^{k-1} \overline{\mathcal{D}}^j(\phi)\partial_{u_j},$$

where $\overline{\mathcal{D}} = \partial_x + u_1\partial_{u_0} + \cdots + f\partial_{u_{k-1}}$ is an operator of a total derivative on $\mathcal{E}$ and $\phi \in C^\infty(\mathcal{E})$ is a generating function.

The generating function is found from the Lie equation

$$\overline{\mathcal{D}}^k(\phi) - \sum_{j=0}^{k-1} f_{u_j}\overline{\mathcal{D}}^j(\phi) = 0. \tag{25}$$

Further on we will use this correspondence by saying that a generating function ϕ is a symmetry.

Consider now the Schrödinger type equations

$$y'' + w(x)y = 0. \tag{26}$$

Equation (26) can be identified with a smooth manifold

$$\mathcal{E} = \{u_2 = -w(x)u_0\} \subset J^2(x, u_0, u_1, u_2). \tag{27}$$

Looking for linear symmetries of (27)

$$\phi = a(x)u_0 + b(x)u_1,$$

where $a(x)$ and $b(x)$ are some functions, we get via (25)

$$\phi = cu_0 + \phi_z, \tag{28}$$

for a constant c, and the function

$$\phi_z = z(x)u_1 - \frac{z'(x)u_0}{2}, \tag{29}$$

satisfying the *Lie equation*:

$$z''' + 4wz' + 2w'z = 0. \tag{30}$$

It is easy to check that if $y(x)$ is a solution to (26), then $z(x) = y^2(x)$ is a solution to (30), and therefore if $y_1(x)$, $y_2(x)$ is a fundamental solution to (26), then $z_1 = y_1^2$, $z_2 = 2y_1y_2$, and $z_3 = y_2^2$ is a fundamental solution to (30).

Equation (30) defines the function $z(x)$, and therefore the symmetry ϕ via (28) and (29). At the same time, (30) can be seen as an equation for the potential $w(x)$, and therefore

$$w(x) = \frac{c_w}{z^2} + \frac{1}{4}\left(\frac{z'}{z}\right)^2 - \frac{z''}{2z}, \tag{31}$$

where c_w is a constant. Moreover, if $\widehat{w}(z)$ is another potential with the same symmetry $z(x)$, then

$$\widehat{w} = w + \frac{\widehat{c}}{z^2},$$

where $\widehat{c} = c_{\widehat{w}} - c_w$ is a constant.

As we will show below, the function $z(x)$ plays an important role in finding fundamental solutions to (26).

We start with

Lemma 3.1 *Function $H = \phi_1\overline{\mathcal{D}}(\phi_2) - \phi_2\overline{\mathcal{D}}(\phi_1)$ is the first integral of* (26) *for any symmetries ϕ_1 and ϕ_2.*

Proof Since ϕ_1 and ϕ_2 are symmetries, $\overline{\mathcal{D}}^2(\phi_{1,2}) = -w(x)\phi_{1,2}$ due to (25).

$$\overline{\mathcal{D}}(H) = \phi_1\overline{\mathcal{D}}^2(\phi_2) - \phi_2\overline{\mathcal{D}}^2(\phi_1) = 0.$$

Putting $\phi_1 = \phi_z$ and $\phi_2 = u_0$, we get the first integral H in the form

$$H = \frac{c_w}{z}(u_0)^2 + \frac{1}{z}(\phi_z)^2.$$

Let now $v = u_0/\sqrt{|z|}$, then $(v')^2 = z^{-3}(\phi_z)^2$, and

$$H = c_w v^2 + z^2\left(v'\right)^2 = H_0^2$$

for some constant $H_0 > 0$. Depending on the sign of the constant c_w we get three different cases

- Elliptic case, $c_w = q_0^2 > 0$.
  Introduce a new variable ψ by the following way:

$$v = \frac{H_0}{q_0} \sin \psi, \quad v' = \frac{H_0}{z} \cos \psi.$$

  The last implies that

$$\psi = \int \frac{q_0}{z} dx, \quad y = \frac{H_0}{q_0} \sqrt{|z|} \sin \left(q_0 \int \frac{dx}{z} \right).$$

- Hyperbolic case, $c_w = -q_0^2 < 0$
  In the same way we obtain

$$y = \frac{H_0}{q_0} \sqrt{|z|} \sinh \left(q_0 \int \frac{dx}{z} \right).$$

- Parabolic case, $c_w = 0$

$$y = H_0 \sqrt{|z|} \int \frac{dx}{z}.$$

Summarizing, we end up with

Theorem 3.12 *Let $z(x)$ be a nonzero symmetry of (26). Then, a fundamental solution of (26) is given as*

- *for $c_w = q_0^2 > 0$*

$$y^{(1)}(x) = \sqrt{|z|} \sin \left(q_0 \int \frac{dx}{z} \right), \quad y^{(2)}(x) = \sqrt{|z|} \cos \left(q_0 \int \frac{dx}{z} \right). \tag{32}$$

- *for $c_w = -q_0^2 < 0$*

$$y^{(1)}(x) = \sqrt{|z|} \sinh \left(q_0 \int \frac{dx}{z} \right), \quad y^{(2)}(x) = \sqrt{|z|} \cosh \left(q_0 \int \frac{dx}{z} \right). \tag{33}$$

- *for $c_w = 0$*

$$y^{(1)}(x) = \sqrt{|z|} \int \frac{dx}{z}, \quad y^{(2)}(x) = \sqrt{|z|}. \tag{34}$$

A fundamental solution of (30) is given as

- *for $c_w = q_0^2 > 0$*

$$z^{(1)}(x) = z, \ z^{(2)}(x) = z \sin \left(2q_0 \int \frac{dx}{z} \right), \ z^{(3)}(x) = z \cos \left(2q_0 \int \frac{dx}{z} \right). \tag{35}$$

• *for* $c_w = -q_0^2 < 0$

$$z^{(1)}(x) = z,\ z^{(2)}(x) = z \sinh\left(2q_0 \int \frac{dx}{z}\right),\ z^{(3)}(x) = z \cosh\left(2q_0 \int \frac{dx}{z}\right). \tag{36}$$

• *for* $c_w = 0$

$$z^{(1)}(x) = z,\ z^{(2)}(x) = z\left(\int \frac{dx}{z}\right)^2,\ z^{(3)}(x) = z \int \frac{dx}{z}. \tag{37}$$

Theorem 3.12 gives us a method of constructing integrable potentials in the following way.

1. Given a pair (z, w)
2. Get a fundamental solution to (26) by means of (32), or (33), or (34)
3. Get a fundamental solution to (30) by means of (35), or (36), or (37)
4. Get a three-parametric family of integrable potentials (for example, in the elliptic case)

$$\widehat{w} = w + \frac{\widehat{c}}{z^2}\left(\alpha_1 + \alpha_2 \sin\left(2q_0 \int \frac{dx}{z}\right) + \alpha_3 \cos\left(2q_0 \int \frac{dx}{z}\right)\right)^{-2}$$

with new symmetries

$$\widehat{z} = z\left(\alpha_1 + \alpha_2 \sin\left(2q_0 \int \frac{dx}{z}\right) + \alpha_3 \cos\left(2q_0 \int \frac{dx}{z}\right)\right),$$

where $\alpha_1, \alpha_2, \alpha_3$ are constants.
5. Again, have a pair $(\widehat{z}, \widehat{w})$ and go to Step 1.

Following the algorithm, we get an infinite chain of potentials $w(x)$, for which the Schrödinger equation (26) is explicitly integrated. In the context of finding solutions to the Euler system (4), this gives us a method of finding infinite chains of barotropic functions $p(\rho)$ for which the quotient Eq. (15) has explicit integrals, and consequently so does (4).

3.4.2 Examples of Solutions

Let us now illustrate how the described method works for concrete choices of pairs (z, w). Recall that we have the following equation for $\nu(\rho)$:

$$\nu'' + w(\rho)\nu = 0,$$

where $w(\rho) = -\lambda A^2(\rho) = -\lambda \rho^{-2} p'(\rho)$.

We start with the pair (z, w) with $c_w = 0$ (parabolic case):

$$z(\rho) = \rho^m, \quad w(\rho) = \frac{m(2-m)}{4\rho^2},$$

where m is a constant.

The next integrable potential $\widehat{w}(\rho)$ is

$$\widehat{w}(\rho) = \frac{m(2-m)}{4\rho^2} + \frac{c_{\widehat{w}}}{\left(\alpha_1\rho^m + \alpha_2\frac{\rho^{2-m}}{(1-m)^2} + \alpha_3\frac{\rho}{1-m}\right)^2}$$

with the symmetry $\widehat{z}(\rho)$

$$\widehat{z}(\rho) = \alpha_1\rho^m + \alpha_2\frac{\rho^{2-m}}{(1-m)^2} + \alpha_3\frac{\rho}{1-m}.$$

We will put $c_{\widehat{w}} = 1$, $\lambda = -1$ and in this case solution to (15) will be

$$f(u, \rho) = (C_1\sin(u) + C_2\cos(u))\left(C_3\sqrt{\widehat{z}}\sin(\xi(\rho)) + C_4\sqrt{\widehat{z}}\cos(\xi(\rho))\right),$$

where

$$\xi(\rho) = \int\frac{d\rho}{z} =$$
$$= \frac{1}{\sqrt{\alpha_3^2 - 4\alpha_1\alpha_2}}\ln\left(\frac{(2\alpha_1\rho^m(m-1) - \rho\alpha_3)\sqrt{\alpha_3^2 - 4\alpha_1\alpha_2} + 4\alpha_1\alpha_2\rho - \alpha_3^2\rho}{(2\alpha_1\rho^m(m-1) - \rho\alpha_3)\sqrt{\alpha_3^2 - 4\alpha_1\alpha_2} - 4\alpha_1\alpha_2\rho + \alpha_3^2\rho}\right).$$

We will proceed with $\alpha_1 = 1$, $\alpha_2 = \alpha_3 = 0$, $m = -2$. In this case the pressure function $p(\rho)$ is

$$p(\rho) = \frac{\rho^7}{7} - 2\rho. \qquad (38)$$

Its graph is shown in Fig. 9 along with that of the van der Waals gas (20). One can observe qualitatively similar behavior, which means that (38) can equally be useful in modeling phase transitions as (20).

In case of $C_2 = C_3 = 0$, $C_4 = 1$ we get a multivalued solution to (4) given by integrals

$$t = C_1\rho^{-2}\sin(u)\cos\left(\frac{\rho^3}{3}\right) + t_0,$$

$$x = C_1\rho^{-2}\left((u\sin(u) - \cos(u))\cos\left(\frac{\rho^3}{3}\right) - \rho^3\cos(u)\sin\left(\frac{\rho^3}{3}\right)\right) + x_0,$$

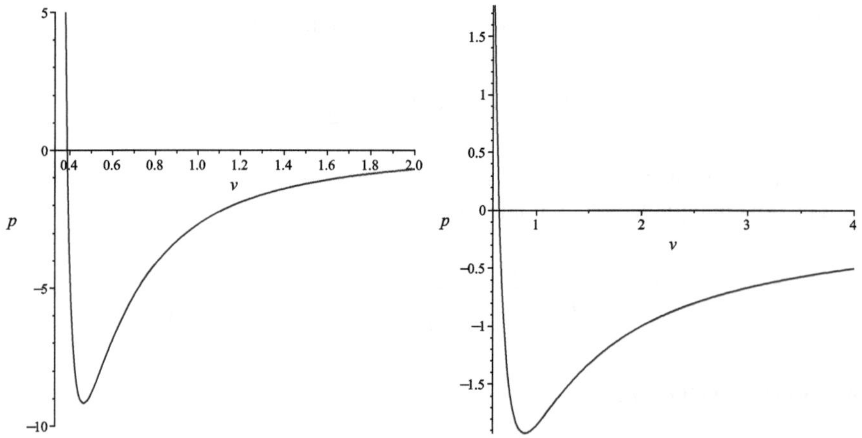

Fig. 9 Dependence of pressure on density for the van der Waals gas (left) and for (38) (right)

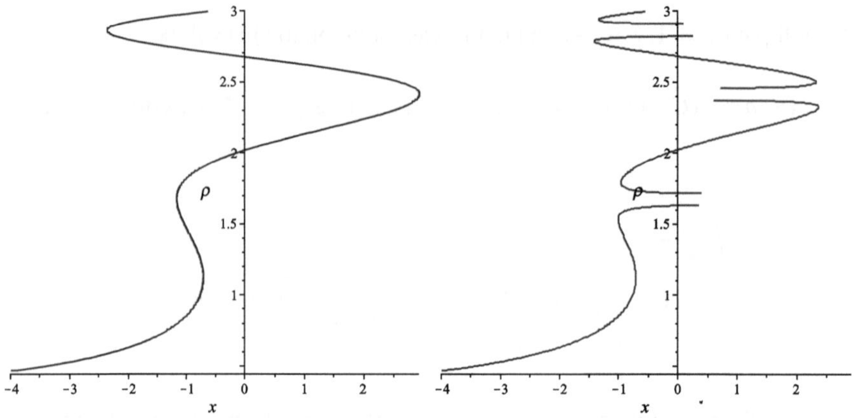

Fig. 10 Distribution of the density over x for two time moments

for some constants x_0, t_0. Its graph is shown in Fig. 10. The behavior shown in Fig. 10 is clearly far from the normal one. However, this illustrates the power of the method for finding multivalued solutions described here, because it is not obvious how to capture such kind of behavior by standard approaches.

4 Barotropic 2D and 3D Euler's Equations

In this section, we will consider the next example of applying the concept of quotient PDEs and differential invariants to find multivalued solutions of barotropic Euler's equations, which is axially and spherically symmetric flows in 2D and 3D, respectively. The system of equations is

$$\mathcal{E} = \begin{cases} \rho_t + r^{-\alpha}(r^{\alpha}\rho u)_r = 0, \\ \rho(u_t + uu_r) + A(\rho)\rho_r = 0, \end{cases} \tag{1}$$

where $\alpha = 1$ corresponds to the axially symmetric flow in 2D, $\alpha = 2$ corresponds to the spherically symmetric flow in 3D, r stands for the radial coordinate in both cases, and $A(\rho) = p'(\rho)$. Since the difference between these two cases is just different values of α, we shall investigate them both simultaneously in a single section.

It is clearly seen from (1) that this system no longer allows for spatial translation symmetry, but still has the scaling symmetry

$$X = t\frac{\partial}{\partial t} + r\frac{\partial}{\partial r}.$$

Let $\mathfrak{g}$ be spanned by X, then the generators of the algebra of $\mathfrak{g}$-invariants are

$$a_1 = u, \quad a_2 = \rho, \quad b^1 = \frac{t}{r}$$

together with Tresse derivatives

$$\frac{d}{du} = \frac{\rho\rho_r r}{u_r\rho^2(\alpha u + ru_r) - Ar\rho_r^2}\frac{d}{dt} + \frac{\rho(ur\rho_r + \rho(\alpha u + ru_r))}{u_r\rho^2(\alpha u + ru_r) - Ar\rho_r^2}\frac{d}{dr},$$
$$\frac{d}{d\rho} = -\frac{\rho ru_r}{u_r\rho^2(\alpha u + ru_r) - Ar\rho_r^2}\frac{d}{dt} - \frac{r(A\rho_r + \rho uu_r)}{u_r\rho^2(\alpha u + ru_r) - Ar\rho_r^2}\frac{d}{dr}.$$

Put $b^1 = t/r = B(u, \rho)$. Finding syzygies among u, ρ, $\frac{dB}{du}$, $\frac{dB}{d\rho}$, $\frac{d^2B}{du^2}$, $\frac{d^2B}{d\rho^2}$, $\frac{d^2B}{dud\rho}$, we get a quotient PDE:

$$KB_{\rho\rho} - 2LB_{u\rho} + MB_{uu} + N = 0,$$

where

$$K = \rho^2\left(A(B - \alpha uB_u)^2 - (Bu - 1)^2\right),$$
$$L = \alpha\rho Au(\alpha\rho uB_\rho B_u - \rho BB_\rho - uBB_u + B_u),$$
$$M = A\left((u(B - \alpha\rho B_\rho) - 1)^2 - AB^2\right),$$

$$N = \rho A_\rho\left(B - \alpha uB_u)(\rho B_\rho(B - \alpha uB_u) + B_u(uB - 1)\right) - (AB_u)^2(\alpha uB_u + (\alpha - 2)B) +$$
$$A\left((\rho B_\rho)^2(\alpha uB_u + B(\alpha - 2)) + \alpha(\alpha + 1)\rho B_\rho(uB_u)^2 - 2B_u(Bu - 1)(uB_u + B)\right) -$$
$$\rho B_\rho\left(\alpha(\alpha + 1)u^2(\rho B_\rho)^2 - 2(\alpha + 1)(Bu - 1)u\rho B_\rho + 2(Bu - 1)^2\right)$$

with the following pair of solution:

$$A(\rho) = \frac{C_1\rho^{1/\alpha}}{(1 + C_2\rho^{1/\alpha})^2}, \quad B(u, \rho) = \frac{1 + C_2\rho^{1/\alpha}}{u}, \tag{2}$$

where C_1, C_2 are constants.

The pressure function $p(\rho)$ corresponding to (2) is

- for $\alpha = 1$

$$p(\rho) = \frac{C_1}{C_2^2}\ln(C_2\rho + 1) + \frac{C_1}{C_2^2(C_2\rho + 1)},$$

- for $\alpha = 2$

$$p(\rho) = \frac{2C_1}{C_2^2}\sqrt{\rho} - \frac{2C_1}{C_2^3(C_2\sqrt{\rho} + 1)} - \frac{4C_1}{C_2^3}\ln\left(C_2\sqrt{\rho} + 1\right).$$

In both cases these functions can be used to describe gases of non-interacting molecules with a finite nonzero volume.

The Cartan distribution $\mathcal{C}_{\widehat{\mathcal{E}}}$ on the overdetermined system

$$\widehat{\mathcal{E}} = \mathcal{E} \cap \left\{\frac{t}{r} - B(u, \rho) = 0,\ \frac{d}{dt}\left(\frac{t}{r} - B(u, \rho)\right) = 0,\ \frac{d}{dr}\left(\frac{t}{r} - B(u, \rho)\right) = 0\right\}$$

is given by the closed differential 1-form

$$\varkappa = \frac{dr}{r} - \frac{C_1\alpha(\alpha + 1) + C_2u^2}{C_1\alpha(\alpha + 1)u}du + \frac{\alpha + C_2(\alpha + 1)\rho^{1/\alpha}}{\alpha(\alpha + 1)\rho\left(1 + C_2\rho^{1/\alpha}\right)}d\rho. \tag{3}$$

Integrating (3) we get the first integral for (1):

$$r = r_0\left(\rho\left(C_2\rho^{1/\alpha} + 1\right)\right)^{-\frac{1}{\alpha+1}} u \exp\left(\frac{C_2u^2}{2C_1\alpha(\alpha + 1)}\right). \tag{4}$$

Using that $t/r = B(u, \rho)$ we get the second integral for (1):

$$t = r_0\rho^{-\frac{1}{\alpha+1}}\left(C_2\rho^{1/\alpha} + 1\right)^{\frac{\alpha}{\alpha+1}} \exp\left(\frac{C_2u^2}{2C_1\alpha(\alpha + 1)}\right) + t_0. \tag{5}$$

Formulas (4)–(5) give a multivalued solution to (1). The distribution of density in case $\alpha = 1$, i.e., the axially symmetric 2D case, is shown in Fig. 11. The corresponding graphs in $3D$ are of a similar form. The choice of constants is $C_1 = 1$, $C_2 = -1$, $t_0 = 1$, $x_0 = -1$.

Again, one can see how an initially smooth solution blows up with the formation of a shock wave.

Further, we will proceed with the analysis of caustics and shock wave fronts for the 2D case $\alpha = 1$. The case $\alpha = 2$ can be addressed in exactly the same manner.

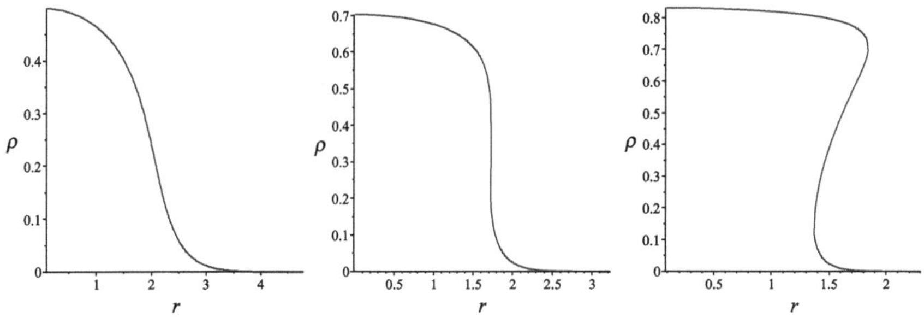

Fig. 11 Density graph for $\alpha = 1$ at moments $t = 0$, $t = 0.35$, $t = 0.55$

Finding equations for the caustic from the condition $dr \wedge dt = 0$ on the solution (4)–(5) we get

$$r(\rho) = \pm \frac{r_0}{C_2\rho}\sqrt{\frac{C_1}{C_2\rho + 1}} \exp\left(\frac{1}{4C_2\rho}\right),$$

$$t(\rho) = t_0 + r_0\sqrt{\frac{C_2\rho + 1}{\rho}} \exp\left(\frac{1}{4C_2\rho}\right).$$

The sign $\pm$ corresponds to positive and negative values of r, and we therefore will be choosing only one of them corresponding to $r > 0$.

The potential for the mass conservation law is

$$H(\rho, t) = -\frac{C_1\rho(t - t_0)^2}{C_2(C_2\rho + 1)^2}\left(C_2\rho \ln\left(\frac{(t - t_0)^2\rho}{r_0^2(C_2\rho + 1)}\right) - C_2\rho - 1\right).$$

The discontinuity curve is found in a similar way as it was done in the previous section. The caustic along with the discontinuity curve, or shock wave front, is shown in Fig. 12.

5 Fully Compressible Euler's Equations

In this section, we address the system of the Euler equations in 1D written for the flows free from the barotropic ansatz, *fully compressible*. The system is

$$\begin{cases} \rho_t + (\rho u)_x = 0, \\ \rho(u_t + uu_x) + p_x = 0, \\ s_t + us_x = 0. \end{cases} \tag{1}$$

Having fixed the thermodynamic state of the medium, system (1) becomes complete.

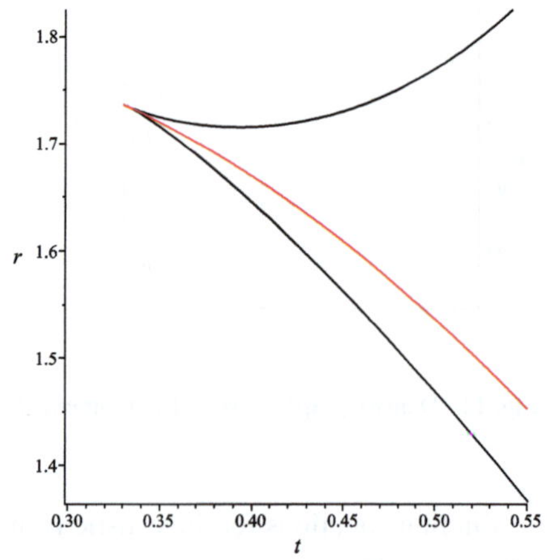

Fig. 12 Caustic (black) and shock wave front (red) for $\alpha = 1$

We will restrict our consideration to the case of ideal gases, and due to (6), where we put $R = 1$ for simplicity, we get the following three-component system

$$\mathcal{E} = \begin{cases} \rho_t + (\rho u)_x = 0, \\ \rho(u_t + uu_x) + p_x = 0, \\ \dfrac{n}{2p}(p_t + up_x) + \left(1 + \dfrac{n}{2}\right) u_x = 0 \end{cases} \tag{2}$$

with unknowns $\rho(t,x)$, $u(t,x)$, $p(t,x)$.

Theorem 3.13 *The symmetry Lie algebra of* (2) *is generated by vector fields*

$$X_1 = \frac{\partial}{\partial x}, \quad X_2 = \frac{\partial}{\partial t}, \quad X_3 = t\frac{\partial}{\partial x} + \frac{\partial}{\partial u}, \quad X_4 = \rho\frac{\partial}{\partial \rho} + p\frac{\partial}{\partial p},$$
$$X_5 = x\frac{\partial}{\partial x} + u\frac{\partial}{\partial u} - 2\rho\frac{\partial}{\partial \rho}, \quad X_6 = t\frac{\partial}{\partial t} - u\frac{\partial}{\partial u} + 2\rho\frac{\partial}{\partial \rho}.$$

Here, we choose invariants of time and space translations, $\mathfrak{g} = \langle X_1, X_2 \rangle$, which are

$$a_1 = u, \quad a_2 = \rho, \quad b^1 = p.$$

As previously, we put $p = B(u, \rho)$.

Before we proceed with the derivation of the quotient PDE, let us note that if we have one constraint on J^0, then the dimension of the overdetermined system $\widehat{\mathcal{E}}$ is $\dim(\widehat{\mathcal{E}}) = 5$, and therefore $\operatorname{codim}(\mathcal{C}_{\widehat{\mathcal{E}}}) = 3$, which means that the distribution $\mathcal{C}_{\widehat{\mathcal{E}}}$ has to have three-dimensional symmetry Lie algebra to be integrated using the Lie-Bianchi theorem. To that end, we additionally say that the function $p - B(u, \rho)$ is invariant with respect to vector field X_4, which implies

$$B(u, \rho) = \rho f(u) \tag{3}$$

for some function $f(u)$ to be found from the quotient PDE. Finding the syzygies among differential $\mathfrak{g}$-invariants and using the assumption (3), we get

$$\left(n(n+2)\left(f'\right)^2 - 4f\right) ff''' - (n+2)f'\left(nf'' - 2\right)\left(2ff'' - \left(f'\right)^2\right) = 0. \quad (4)$$

Solutions to (4) are given implicitly by

$$\int \frac{\alpha_1\sqrt{n(n+2)}}{\sqrt{2f\left(\alpha_2 f^{\alpha_1^2} - 2(n+1-\alpha_1^2)\right)}} df = u + \alpha_3,$$

where α_j are constants.

Now that we have solution to (4), we get a completely integrable distribution $\mathcal{C}_{\widehat{\mathcal{E}}}$, where

$$\widehat{\mathcal{E}} = \mathcal{E} \cap \left\{p - \rho f(u) = 0,\ \frac{d}{dt}\left(p - \rho f(u)\right) = 0,\ \frac{d}{dx}\left(p - \rho f(u)\right) = 0\right\},$$

of codimension 3 with the 3-dimensional commutative symmetry Lie algebra $\mathfrak{g} = \langle X_1, X_2, X_4 \rangle$.

To proceed, we describe the algorithm of finding integrals of a completely integrable distribution with the commutative symmetry Lie algebra [28].

5.1 Lie-Bianchi Theorem

The following result known as the Lie-Bianchi theorem shall be important for us.

Theorem 3.14 (Lie-Bianchi) *Let P be a completely integrable m-codimensional distribution with the solvable m-dimensional symmetry Lie algebra $\mathfrak{g}$ transversal to P. Then, P is integrable by quadratures.*

We will restrict to the discussion of the case when $\mathfrak{g}$ is commutative. For the general case we refer to [28].

Let P be a completely integrable distribution of codimension m generated by differential 1-forms $\omega_1, \dots, \omega_m$. Let $\mathfrak{g}$ be the Lie algebra of its shuffle symmetries generated by vector fields $X_1, \dots, X_m$. Then, the matrix

$$W = \left\| \omega_i(X_j) \right\|_{i,j=1,\dots,m}$$

is non-degenerate, and a new basis $\overline{\omega}_1, \dots, \overline{\omega}_m$ given by

$$\begin{pmatrix} \overline{\omega}_1 \\ \vdots \\ \overline{\omega}_m \end{pmatrix} = W^{-1} \begin{pmatrix} \omega_1 \\ \vdots \\ \omega_m \end{pmatrix} \quad (5)$$

satisfies the condition

$$\overline{\omega}_i(X_j) = \delta_{ij}. \tag{6}$$

Let now c_{ij}^l be the structure constants of the Lie algebra $\mathfrak{g}$:

$$[X_i, X_j] = \sum_{l=1}^{m} c_{ij}^l X_l.$$

Lemma 3.2 *([28]) Let P be a completely integrable distribution defined by differential 1-forms $\omega_1, \ldots, \omega_m$ that satisfy* (6)*, and let c_{ij}^l be the structure constants of the symmetry Lie algebra $\mathfrak{g}$. Then,*

$$d\omega_l + \sum_{i<j} c_{ij}^l \omega_i \wedge \omega_j = 0. \tag{7}$$

Let now the Lie algebra $\mathfrak{g}$ be commutative, i.e., $c_{ij}^l = 0$ for all $i, j, l = 1, \ldots, m$. Then, due to (7), $d\omega_l = 0, l = 1, \ldots, m$, and integrals of P are

$$H_i = \int_\Gamma \omega_i, \quad i = 1, \ldots, m.$$

Remark 3.2 If $\mathrm{codim}(P) = 1$, P is generated by the differential form ω, and X is a transversal symmetry of P, i.e., $\omega(X) \neq 0$, then the integral of P is

$$H = \int_\Gamma \frac{\omega}{\omega(X)}.$$

We used this for integration of completely integrable 1-codimensional Cartan distributions in previous sections for barotropic Euler's equations.

We will proceed with finding solutions to (2) for two cases, $\alpha_2 = 0$, and $\alpha_1 = 1/\sqrt{2}$, which we can easily illustrate.

5.2 *The Case $\alpha_2 = 0$*

Here, solution to (4) can be written as

$$f(u) = C_0(u + \alpha_3)^2,$$

where

$$C_0 = \frac{\alpha_1^2 - n - 1}{\alpha_1^2 n(n+2)}.$$

Let (t, x, u, ρ, ρ_t) be the local coordinates on $\widehat{\mathcal{E}}$. Restriction of the Cartan forms

$$\omega_1 = du - u_x dx - u_t dt,$$
$$\omega_2 = d\rho - \rho_x dx - \rho_t dt,$$
$$\omega_3 = d\rho_t - \rho_{tx} dx - \rho_{tt} dt$$

to $\widehat{\mathcal{E}}$ determines completely integrable distribution $\mathcal{C}_{\widehat{\mathcal{E}}}$ with the commutative symmetry Lie algebra $\mathfrak{g} = \langle X_1, X_2, X_4 \rangle$. Constructing the new forms according to (5), we get

$$\overline{\omega}_1 = dt - \frac{\alpha_3 \rho (2C_0 n - 1)}{C_0(n+1)\rho_t (u+\alpha_3)^2} du + \frac{u - C_0 n(u - \alpha_3)}{C_0(n+1)\rho_t(u+\alpha_3)} d\rho - \frac{\rho(nC_0(\alpha_3 - u) + u)}{C_0(n+1)(u+\alpha_3)\rho_t^2} d\rho_t,$$

$$\overline{\omega}_2 = dx - \frac{\rho\left(C_0 n^2\left(u^2 + 2u\alpha_3 - \alpha_3^2\right) + n\left(C_0(u+\alpha_3)^2 - u(u+2\alpha_3)\right) - (u+\alpha_3)^2\right)}{C_0 n(n+1)\rho_t(u+\alpha_3)^2} du -$$
$$\frac{(u(n+1)+\alpha_3)(C_0 n(u-\alpha_3) - u)}{C_0 n(n+1)\rho_t(u+\alpha_3)} d\rho - \frac{(u(n+1)+\alpha_3)(nC_0(\alpha_3 - u) + u)\rho}{C_0 n(n+1)(u+\alpha_3)\rho_t^2} d\rho_t,$$

$$\overline{\omega}_3 = \frac{(C_0 n - 1)(u(n+1)+\alpha_3)(2C_0 n - 1)}{C_0 n(n+1)(u+\alpha_3)(C_0 n(u-\alpha_3) - u)} du +$$
$$\frac{2C_0 n^2 + 3C_0 n - 1}{C_0 n(n+1)\rho} d\rho - \frac{C_0 n^2 + 2C_0 n - 1}{C_0 n(n+1)\rho_t} d\rho_t.$$

One can check that 1-forms $\overline{\omega}_1, \overline{\omega}_2, \overline{\omega}_3$ are closed, and their integrals are

$$t = \frac{\rho(C_0 n(u-\alpha_3) - u)}{C_0(n+1)(u+\alpha_3)\rho_t} + t_0, \tag{8}$$

$$x = \frac{\rho(u(n+1)+\alpha_3)(C_0 n(u-\alpha_3) - u)}{C_0 n(n+1)(u+\alpha_3)\rho_t} + x_0, \tag{9}$$

where x_0, t_0 are constants, and ρ_t is found from the formula

$$\rho_t = C_1(C_0 n(u-\alpha_3) - u)\left((u+\alpha_3)^{n(1-C_0 n)}\rho^{1-C_0 n(2n+3)}\right)^{-\frac{1}{C_0 n(n+2)-1}}. \tag{10}$$

Formulas (8)–(10) give us solution N for the density ρ and the velocity u implicitly, while the pressure p is found from

$$p = C_0 \rho (u+\alpha_3)^2.$$

Restriction of the differential 2-form $dt \wedge dx$ to the solution (8)–(10) gives

$$(dt \wedge dx)|_N = \frac{\rho\left((u+\alpha_3)^{-n(C_0 n-1)}\rho^{1-C_0 n(2n+3)}\right)^{\frac{2}{C_0 n(n+2)-1}}}{C_1^2 C_0 (u+\alpha_3)^2 (C_0 n(n+2)-1)} du \wedge d\rho,$$

and we do not observe shock wave type singularities.

5.3 The Case $\alpha_1 = 1/\sqrt{2}$

Here, solution to (4) can be written as

$$f(u) = \frac{\left(n^3 + \frac{5n^2}{2} + n + \alpha_2(u+\alpha_3)^2\right)^2}{2\alpha_2 n^2 (n+2)^2}.$$

Following exactly the same steps as in the previous case, we get integrals for (4) in the form

$$t = t_0 - \frac{n\alpha_2 \rho^{-\frac{2n+1}{2n+4}} (u+\alpha_3)\left(2\alpha_2(u+\alpha_3)^2 + 2n^3 + 5n^2 + 2n\right)^{-\frac{3n}{2n+4}}}{\left(n+\frac{1}{2}\right)\left(n^3 + \frac{5n^2}{2} + n + \alpha_2(u+\alpha_3)^2\right)},$$

$$x = x_0 - \frac{\rho^{-\frac{2n+1}{2n+4}}\left(n^3 + \frac{5n^2}{2} + n(1 + 2\alpha_2 u(u+\alpha_3)) + \alpha_2(u+\alpha_3)^2\right)}{(2n+1)\left(n^3 + \frac{5n^2}{2} + n + \alpha_2(u+\alpha_3)^2\right)} \cdot \left(2\alpha_2(u+\alpha_3)^2 + 2n^3 + 5n^2 + 2n\right)^{-\frac{3n}{2n+4}},$$

where x_0, t_0 are constants. Simpler, the velocity $u(t,x)$ is given implicitly as

$$\frac{x - x_0}{t - t_0} = \frac{2\alpha_2(u+\alpha_3)(u(2n+1)+\alpha_3) + 2n^3 + 5n^2 + 2n}{4n\alpha_2(u+\alpha_3)}.$$

Therefore, singularities appear at

$$u_{1,2} = \frac{-2\alpha_2\alpha_3 \pm \sqrt{2\alpha_2 n(n+2)}}{2\alpha_2}.$$

Graph for the velocity u is given in Fig. 13, where we use $t_0 = 1, x_0 = 0, \alpha_2 = \alpha_3 = 1$, $n = 3$.

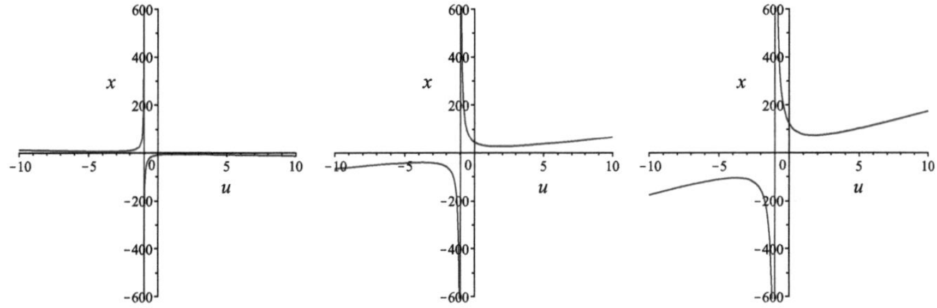

Fig. 13 Velocity graph at moments $t = 0$, $t = 6.25$, $t = 15$

Appendix

Many computations for this paper have been performed in Maple with the Differential Geometry package [52]. Here, we give the codes for essential computations involving differential geometry objects, when using the Maple programming language makes it more efficient and simpler.

Computations for Sect. 3.1.

```
restart;
```

Necessary libraries

```
with(DifferentialGeometry):with(JetCalculus):with(PDETools):
with(Tools):with(LinearAlgebra):
Preferences("JetNotation","JetNotation2"):
DGsetup([t,x],[u,rho],M,0):
declare(u(t,x),rho(t,x),p(rho),A(rho));
```

Forms for the Euler equations

```
omega1:=evalDG(drho &w dx-u*drho &w dt-rho*du &w dt):
omega2:=evalDG(du &w dx-u*du &w dt-diff(p(rho),rho)/rho*drho &w dt):
```

Operator P_ω

```
Omega11:=evalDG(omega1 &w omega1):
w11:=op(DGinfo(Omega11,"CoefficientList","all")):
Omega12:=evalDG(omega1 &w omega2):
w12:=op(DGinfo(Omega12,"CoefficientList","all")):
Omega21:=evalDG(omega2 &w omega1):
w21:=op(DGinfo(Omega21,"CoefficientList","all")):
Omega22:=evalDG(omega2 &w omega2):
w22:=op(DGinfo(Omega22,"CoefficientList","all")):
W:=Matrix([[w11,w12],[w21,w22]]):
```

Effective forms

```
kappa1:=evalDG(a11*omega1+a12*omega2):
kappa2:=evalDG(a21*omega1+a22*omega2):
```

Effectivity conditions

```
eq1:=evalDG(kappa1 &w kappa2):
eq2:=simplify(evalDG(kappa1 &w kappa1+kappa2 &w kappa2)):
req1:=op(DGinfo(eq1,"CoefficientList","all")):
req2:=op(DGinfo(eq2,"CoefficientList","all")):
```

Solving equations for the coefficients

```
solve([req1,req2],{a11,a12,a21,a22}):
suba:={a11=A(rho),a22=1,a12=0,a21=0}:
subp:={diff(p(rho),rho)=rho^2*A(rho)^2}:
kappa1:=eval(eval(kappa1,suba),subp):
kappa2:=eval(eval(kappa2,suba),subp):
```

Matrix of the operator A_ω

```
ss1:=solve(DGinfo(evalDG(Hook(D_t,kappa2)-
Hook(q1*D_t+q2*D_x+q3*D_u+q4*D_rho,kappa1)),
"CoefficientList", "all"),{q1,q2,q3,q4}):
m11:=rhs(ss1[1]):
m21:=rhs(ss1[2]):
m31:=rhs(ss1[3]):
m41:=rhs(ss1[4]):
ss2:=solve(DGinfo(evalDG(Hook(D_x,kappa2)-
Hook(q1*D_t+q2*D_x+q3*D_u+q4*D_rho,kappa1)),
"CoefficientList", "all"),{q1,q2,q3,q4}):
m12:=rhs(ss2[1]):
m22:=rhs(ss2[2]):
m32:=rhs(ss2[3]):
m42:=rhs(ss2[4]):
ss3:=solve(DGinfo(evalDG(Hook(D_u,kappa2)-
Hook(q1*D_t+q2*D_x+q3*D_u+q4*D_rho,kappa1)),
"CoefficientList", "all"),{q1,q2,q3,q4}):
m13:=rhs(ss3[1]):
m23:=rhs(ss3[2]):
m33:=rhs(ss3[3]):
m43:=rhs(ss3[4]):
ss4:=solve(DGinfo(evalDG(Hook(D_rho,kappa2)-
Hook(q1*D_t+q2*D_x+q3*D_u+q4*D_rho,kappa1)),
"CoefficientList", "all"),{q1,q2,q3,q4}):
m14:=rhs(ss4[1]):
m24:=rhs(ss4[2]):
m34:=rhs(ss4[3]):
m44:=rhs(ss4[4]):
```

```
Mm:=simplify(Matrix([[m11,m12,m13,m14],[m21,m22,m23,m24],
[m31,m32,m33,m34],[m41,m42,m43,m44]])):
```

Eigenvalues and eigenvectors of the operator A_ω

```
Eigenvalues(Mm):
Eigenvectors(Mm):
```

Distribution $\mathcal{C}_+$

```
X1:=evalDG(A(rho)*D_u+D_rho):
X2:=evalDG(1/(-A(rho)*rho+u)*D_t+D_x):
```

Distribution $\mathcal{C}_-$

```
Y1:=evalDG(-A(rho)*D_u+D_rho):
Y2:=evalDG(1/(A(rho)*rho+u)*D_t+D_x):
```

Lie bracket for $\mathcal{C}_\pm$

```
X3:=LieBracket(X1,X2):
Y3:=LieBracket(Y1,Y2):
```

Finding the vector fields $V_\pm$

```
equa1:=Hook(evalDG(alpha1*X1+alpha2*X2),
ExteriorDerivative(t-f(u,rho))):
equa2:=Hook(evalDG(beta1*Y1+beta2*Y2),
ExteriorDerivative(t-f(u,rho))):
subconst:=solve(eval([equa1,equa2],{alpha1=1,beta1=1}),
{alpha2,beta2}):
V1:=eval(evalDG(X1+alpha2*X2),subconst):
V2:=eval(evalDG(Y1+beta2*Y2),subconst):
```

Finding the integrability condition $V_- \wedge V_+ \wedge [V_-, V_+] = 0$

```
Vbr:=LieBracket(V1,V2):
triVec:=evalDG(V1 &w V2 &w Vbr):
cf:=DGinfo(triVec,"CoefficientList","all");
```

Computations for Sect. 3.2

```
restart;
```

Necessary libraries

```
with(DifferentialGeometry): with(JetCalculus):with(PDEtools):
Preferences("JetNotation","JetNotation2"):
DGsetup([t, x], [u, rho], M, 2):
```

```
declare(A(rho[0,0]));
```

Euler's equation and its prolongation

```
jeq1:=rho[0,0]*A(rho[0,0])^2*rho[0,1]+u[0,0]*u[0,1]+u[1,0]:
jeq1t:=TotalDiff(jeq1,t):
jeq1x:=TotalDiff(jeq1,x):
```

Continuity equation and its prolongation

```
jeq2:=rho[1,0]+rho[0,1]*u[0,0]+rho[0,0]*u[0,1]:
jeq2t:=TotalDiff(jeq2,t):
jeq2x:=TotalDiff(jeq2,x):
```

Restriction to the Euler system

```
subeq:=eliminate([jeq1,jeq2,jeq1t,jeq1x,jeq2t,jeq2x],
{u[1,0],rho[1,0],rho[2,0],rho[1,1],u[2,0],u[1,1]})[1]:
```

x-translation invariants

```
X:=u[0,0]:
Y:=rho[0,0]:
H:=t:
```

Equations for the coefficients of Tresse derivatives

```
eq1:=a1*TotalDiff(X,t)+a2*TotalDiff(X,x)-1:
eq2:=a1*TotalDiff(Y,t)+a2*TotalDiff(Y,x):
eq3:=a3*TotalDiff(X,t)+a4*TotalDiff(X,x):
eq4:=a3*TotalDiff(Y,t)+a4*TotalDiff(Y,x)-1:
```

Solution for the coefficients

```
suba:=simplify(solve(eval([eq1,eq2,eq3,eq4],subeq),
{a1,a2,a3,a4}),size):
```

Tresse Derivatives

```
TrDerX:=proc(f)
local tdX;
tdX:=simplify(eval(eval(a1*TotalDiff(f,t)
+a2*TotalDiff(f,x),suba),subeq),size);
end proc:
TrDerY:=proc(f)
local tdY;
tdY:=simplify(eval(eval(a3*TotalDiff(f,t)
+a4*TotalDiff(f,x),suba),subeq),size);
end proc:
```

Finding syzygies among differential invariants

```
invs:=eval([TrDerX(TrDerX(H))-B[uu],TrDerY(TrDerX(H))-B[ur],
TrDerY(TrDerY(H))-B[rr], TrDerX(H)-B[u],TrDerY(H)-B[r],
H-f,X-u,Y-r],subeq):
syz1:=eliminate(invs,{rho[0,2],u[0,2],u[0,0],rho[0,0],t})[2]:
syz2:=eliminate(syz1,{u[0,1]})[2]:
subrho01:={rho[0,1]=solve(syz2[1]/rho[0,1],rho[0,1])}:
quoPDE:=simplify(eval(syz2,subrho01),size);
```

Computations for Sect. 5

```
restart;
```

Necessary libraries

```
with(DifferentialGeometry):with(JetCalculus):with(PDETools):
with(Tools):with(LinearAlgebra):with(ListTools):
with(LieAlgebras):Preferences("JetNotation","JetNotation2"):
DGsetup([t, x], [u, rho, p], M, 3):
declare(S(p[0,0],rho[0,0]),rho(t,x),u(t,x),p(t,x),f(u[0,0])):
```

Defining the equations

```
S(p[0,0],rho[0,0]):=ln(p[0,0]^(n/2)/rho[0,0]^(n/2+1)):
eq1:=diff(p(t,x),x)+
rho(t,x)*(u(t,x)*diff(u(t,x),x)+diff(u(t,x),t)):
eq2:=diff(rho(t,x),t)+
u(t,x)*diff(rho(t,x),x)+rho(t,x)*diff(u(t,x),x):
jeq3:=diff(S(p[0,0],rho[0,0]),p[0,0])*(p[1,0]+u[0,0]*p[0,1])-
diff(S(p[0,0],rho[0,0]),rho[0,0])*rho[0,0]*u[0,1]:
eq3:=convert(jeq3,DGdiff):
```

Computing the symmetries

```
syms:=[Infinitesimals([eq1,eq2,eq3])]:
symsV:=convert(syms,DGvector):
```

Prolongation of equations

```
jeq1:=convert(eq1,DGjet):
jeq1t:=TotalDiff(jeq1,t):
jeq1x:=TotalDiff(jeq1,x):
jeq1tt:=TotalDiff(jeq1t,t):
jeq1xx:=TotalDiff(jeq1x,x):
jeq1tx:=TotalDiff(jeq1t,x):
jeq2:=convert(eq2,DGjet):
jeq2t:=TotalDiff(jeq2,t):
jeq2x:=TotalDiff(jeq2,x):
```

```
jeq2tt:=TotalDiff(jeq2t,t):
jeq2xx:=TotalDiff(jeq2x,x):
jeq2tx:=TotalDiff(jeq2t,x):
jeq3t:=TotalDiff(jeq3,t):
jeq3x:=TotalDiff(jeq3,x):
jeq3tt:=TotalDiff(jeq3t,t):
jeq3xx:=TotalDiff(jeq3x,x):
jeq3tx:=TotalDiff(jeq3t,x):
```

Adding the constraint and its prolongation

```
jeq0:=p[0,0]-f(u[0,0])*rho[0,0]:
jeq0t:=TotalDiff(jeq0,t):
jeq0x:=TotalDiff(jeq0,x):
jeq0tt:=TotalDiff(jeq0t,t):
jeq0xx:=TotalDiff(jeq0x,x):
jeq0tx:=TotalDiff(jeq0t,x):
jeq0ttt:=TotalDiff(jeq0tt,t):
jeq0xxt:=TotalDiff(jeq0xx,t):
jeq0txt:=TotalDiff(jeq0tx,t):
jeq0xxx:=TotalDiff(jeq0xx,x):
```

Finding the quotient equation

```
j0:=[jeq0]:
j1:=[jeq1,jeq2,jeq3,jeq0t,jeq0x]:
j2:=[jeq1t,jeq1x,jeq2t,jeq2x,jeq3t,jeq3x,jeq0tt,jeq0tx,jeq0xx]:
j3:=[jeq1tt,jeq1xx,jeq1tx,jeq2tt,jeq2xx,jeq2tx,jeq3tt,jeq3xx,
jeq3tx,jeq0ttt,jeq0txt,jeq0xxx,jeq0xxt]:
el1:=eliminate(j3,{u[3, 0], u[2, 1], u[1, 2], u[0, 3], rho[3, 0],
rho[2, 1], rho[1, 2], rho[0, 3], p[3, 0], p[2, 1], p[1, 2],
p[0, 3]}):
elr1:=el1[2]:
j22:=Flatten([op(elr1),j2]):
el2:=eliminate(j22,{u[2, 0], u[1, 1], u[0, 2], rho[2, 0],
rho[1, 1], rho[0, 2], p[2, 0], p[1, 1], p[0, 2]}):
elr2:=el2[2]:
j11:=Flatten([op(elr2),j1]):
el3:=eliminate(j11,{u[1, 0], u[0, 1], rho[0, 1], p[1, 0],
p[0, 1]}):
elr3:=el3[2]:
j00:=Flatten([op(elr3),j0]):
el4:=eliminate(j00,{p[0,0]}):
elr4:=el4[2]:
quoPDE:=factor(elr4[1]/rho[1,0]):
```

Specifying solution to the quotient equation

```
dsolve(quoPDE):
subf:={f(u[0,0])=C0*(u[0,0]+alpha[3])^(2)}:
```

Restriction to the overdetermined system

```
subE:=simplify(eval(Flatten([op(simplify(eval(el3[1],el4[1]),
size)),p[0, 0] = f(u[0,0])*rho[0,0]]),subf),power):
```

Cartan forms

```
omega1:=evalDG(du[0,0]-u[1,0]*dt-u[0,1]*dx):
omega2:=evalDG(drho[0,0]-rho[1,0]*dt-rho[0,1]*dx):
omega3:=evalDG(drho[1,0]-rho[2,0]*dt-rho[1,1]*dx):
```

Restricting the Cartan forms to the system of equations

```
CartForm:=simplify(eval(eval(eval([omega1,omega2,omega3],
el2[1]),subE),subf),size):
```

The Lie algebra structure for $\mathfrak{g} = \langle X_1, X_2, X_4 \rangle$

```
V1:=Prolong(symsV[1],1):
V2:=Prolong(symsV[2],1):
V4:=Prolong(symsV[4],1):
LieAlgebraData([V1,V2,V4]):
```

Finding the matrix W and its inverse

```
w11:=Hook(V1,CartForm[1]):
w12:=Hook(V2,CartForm[1]):
w13:=Hook(V4,CartForm[1]):
w21:=Hook(V1,CartForm[2]):
w22:=Hook(V2,CartForm[2]):
w23:=Hook(V4,CartForm[2]):
w31:=Hook(V1,CartForm[3]):
w32:=Hook(V2,CartForm[3]):
w33:=Hook(V4,CartForm[3]):
W:=simplify(eval(Matrix([[w11,w12,w13],[w21,w22,w23],
[w31,w32,w33]]),subE),size):
W1:=simplify(MatrixInverse(W)):
W11:=simplify(W1[1][1],size):
W12:=simplify(W1[1][2],size):
W13:=simplify(W1[1][3],size):
W21:=simplify(W1[2][1],size):
W22:=simplify(W1[2][2],size):
W23:=simplify(W1[2][3],size):
W31:=simplify(W1[3][1],size):
W32:=simplify(W1[3][2],size):
W33:=simplify(W1[3][3],size):
```

Finding the new forms

```
kappa1:=evalDG(simplify(evalDG(W11*CartForm[1]+W12*CartForm[2]+
W13*CartForm[3]),size)):
kappa2:=simplify(evalDG(W21*CartForm[1]+W22*CartForm[2]+
```

```
W23*CartForm[3]),size):
kappa3:=simplify(evalDG(W31*CartForm[1]+W32*CartForm[2]+
W33*CartForm[3]),size):
```

Check that the new forms are closed

```
simplify(ExteriorDerivative(kappa1),size):
simplify(ExteriorDerivative(kappa2),size):
simplify(ExteriorDerivative(kappa3),size):
```

Finding the integrals

```
cf1:=DGinfo(kappa1,"CoefficientList", "all"):
cf2:=DGinfo(kappa2,"CoefficientList", "all"):
cf3:=DGinfo(kappa3,"CoefficientList", "all"):
declare(J1(t,u[0,0],rho[0,0],rho[1,0]),
J2(x,u[0,0],rho[0,0],rho[1,0]),J3(u[0,0],rho[0,0],rho[1,0]));
sysJ1:=[diff(J1(t,u[0,0],rho[0,0],rho[1,0]),t)=cf1[1],
diff(J1(t,u[0,0],rho[0,0],rho[1,0]),u[0,0])=cf1[2],
diff(J1(t,u[0,0],rho[0,0],rho[1,0]),rho[0,0])=cf1[3],
diff(J1(t,u[0,0],rho[0,0],rho[1,0]),rho[1,0])=cf1[4]]:
sol1:=pdsolve(sysJ1):
sysJ2:=[diff(J2(x,u[0,0],rho[0,0],rho[1,0]),x)=cf2[1],
diff(J2(x,u[0,0],rho[0,0],rho[1,0]),u[0,0])=cf2[2],
diff(J2(x,u[0,0],rho[0,0],rho[1,0]),rho[0,0])=cf2[3],
diff(J2(x,u[0,0],rho[0,0],rho[1,0]),rho[1,0])=cf2[4]]:
sol2:=pdsolve(sysJ2):
sysJ3:=[diff(J3(u[0,0],rho[0,0],rho[1,0]),u[0,0])=cf3[1],
diff(J3(u[0,0],rho[0,0],rho[1,0]),rho[0,0])=cf3[2],
diff(J3(u[0,0],rho[0,0],rho[1,0]),rho[1,0])=cf3[3]]:
sol3:=simplify(pdsolve(sysJ3),size):
```

References

1. M. Roop, Singularities in one-dimensional Euler flows. J. Geom. Phys. **166**, 104272 (2021)
2. V. Lychagin, M. Roop, Shock waves in Euler flows of gases. Lobachevskii J. Math. **41**(12), 2466–2472 (2020)
3. V. Lychagin, M. Roop, Singularities in Euler flows: multivalued solutions. Shock., Phase Trans., Symmetry **13**, 54 (2021)
4. G.K. Batchelor, *An Introduction to Fluid Dynamics* (Cambridge University Press, Cambridge, 2000)
5. L.D. Landau, E.M. Lifshitz, Fluid mechanics, in *Course of Theoretical Physics*, vol. 6. (Pergamon Press, Oxford, 1987)
6. A.E. Scheidegger, The physics of flow through porous media, in *Revised edition* (The Macmillan Co., New-York, 1960)
7. L.S. Leibenson, Motion of natural liquids and gases in a porous medium (Gostekhizdat, Moscow, 1947)
8. M. Muskat, *The Flow of Homogeneous Fluids Through Porous Media* (McGraw-Hill, New-York, 1937)

9. H. Henk, *Fluids in Porous Media: Transport and Phase Changes* (Morgan & Claypool Publishers. 2016), 89 p.
10. R. Masoodi, K.M. Pillai, *Wicking in Porous Materials: Traditional and Modern Modeling Approaches* (CRC Press Taylor & Francis Group, New York, 2013), p.380
11. V. Fortov, *Thermodynamics and Equations of State for Matter: From Ideal Gas to Quark-Gluon Plasma* (World Scientific Publishing, 2016)
12. J.W. Gibbs, A method of geometrical representation of the thermodynamic properties of substances by means of surfaces, in *Transactions of the Connecticut Academy*, vol. 1 (1873), pp. 382–404
13. V. Lychagin, Contact geometry, measurement, and thermodynamics, in *Nonlinear PDEs, Their Geometry and Applications*. ed. by R. Kycia, E. Schneider, M. Ulan (Switzerland, Birkhäuser, Cham, 2019), pp.3–52
14. V.I. Arnold, *Singularities of Caustics and Wave Fronts* (Springer, Netherlands, 1990)
15. V.I. Arnold, *Catastrophe Theory* (Springer, Berlin, Heidelberg, 1984)
16. V. Arnold, S. Gusein-Zade, A. Varchenko, *Singularities of Differentiable Maps* (Birkhäuser, Basel, 1985)
17. V. Lychagin, M. Roop, Phase transitions in filtration of Redlich-Kwong gases. J. Geom. Phys. **143**, 33–40 (2019)
18. V. Lychagin, M. Roop, Critical phenomena in filtration processes of real gases. Lobachevskii J. Math. **41**(3), 382–399 (2020)
19. V. Lychagin, M. Roop, Real gas flows issued from a source. Anal. Math. Phys. **10**(1), 3 (1–16) (2020)
20. V. Lychagin, M. Roop, Critical phenomena in Darcy and Euler flows of real gases, in E. Schneider, M. Ulan (eds.) *Differential Geometry, Differential Equations, and Mathematical Physics* (Springer Nature, Basel, Switzerland, 2021), pp. 151–186
21. S.J. Huang, R. Wang, On blowup phenomena of solutions to the Euler equations for Chaplygin gases. Appl. Math. Comput. **219**, 4365–4370 (2013)
22. R. Rosales, E. Tabak, Caustics of weak shock waves. Phys. Fluids **10**(1), 206–222 (1997)
23. R. Chaturvedi, P. Gupta, L.P. Singh, Evolution of weak shock wave in two-dimensional steady supersonic flow in dusty gas. Acta Astronaut. **160**, 552–557 (2019)
24. A. Vinogradov, I. Krasilshchik (eds.), *Symmetries and Conservation Laws for Differential Equations of Mathematical Physics* (Factorial, Moscow, 1997)
25. A. Vinogradov, I. Krasilshchik, V. Lychagin, *Geometry of Jet Spaces and Nonlinear Partial Differential Equations* (Gordon and Breach, New York, 1996)
26. L. Ovsiannikov, *Group Analysis of Differential Equations* (Academic Press, 1982)
27. P. Olver, *Applications of Lie Groups to Differential Equations* (Springer, New York, 1986)
28. A. Kushner, V. Lychagin, V. Rubtsov, *Contact Geometry and Nonlinear Differential Equations* (Cambridge University Press, Cambridge, 2007)
29. V. Lychagin, Contact geometry and non-linear second order differential equations. Russ. Math. Surv. **34**(1), 149–180 (1979)
30. B. Banos, J. Gibbon, I. Roulstone, V. Rubtsov, Kähler geometry and the Navier-Stokes equations (2005). arXiv:nlin/0509023
31. B. Banos, I. Roulstone, V. Rubtsov, Monge-Ampère structures and the geometry of incompressible flows. J. Phys. A: Math. Theor. **49**(24) (2016). https://doi.org/10.1088/1751-8113/49/24/244003.
32. R. D'Onofrio, G. Ortenzi, I. Roulstone, V. Rubtsov, Solutions and singularities of the semigeostrophic equations via the geometry of Lagrangian submanifolds (2022). arXiv:2209.13337
33. D. Tunitsky, On the global solubility of the Cauchy problem for hyperbolic Monge-Ampère systems. Izv.: Math. **82**(5), 1019–1075 (2018)
34. D. Tunitsky, On global solvability of initial value problem for hyperbolic Monge-Ampère equations and systems. Dokl. Math. **96**(1), 1–3 (2017)
35. D. Tunitsky, On multivalued solutions of equations of one-dimensional gas flow, in *Proceedings of the 12th International Conference "Management of Large-Scale System Development" (MLSD)* (2019), pp. 1–3

36. D. Tunitsky, I. Bogaevsky, Singularities of multivalued solutions of quasilinear hyperbolic systems. Proc. Steklov Inst. Math. **308**, 67–78 (2020)
37. V. Lychagin, Singularities of multivalued solutions of nonlinear differential equations, and nonlinear phenomena. Acta Appl. Math. **3**(2), 135–173 (1985)
38. A. Akhmetzyanov, A. Kushner, V. Lychagin, Control of displacement front in a model of immiscible two-phase flow in porous media. Dokl. Math. **94**(1), 378–381 (2016)
39. A. Akhmetzyanov, A. Kushner, V. Lychagin, Integrability of Buckley-Leverett's filtration model. IFAC-PapersOnLine **49**(12), 1251–1254 (2016)
40. A. Akhmetzyanov, A. Kushner, V. Lychagin, Shock waves in initial boundary value problem for filtration in two-phase 2-dimensional porous media. Glob. Stoch. Anal. **3**(2), 41–46 (2016)
41. B. Kruglikov, V. Lychagin, Compatibility, multi-brackets and integrability of systems of PDEs. Acta Appl. Math. **109**, 151–196 (2010)
42. B. Kruglikov, V. Lychagin, Global Lie-Tresse theorem. Selecta Math. **22**, 1357–1411 (2016)
43. V. Lychagin, V. Yumaguzhin, On geometric structures of 2-dimensional gas dynamics equations. Lobachevskii J. Math. **30**(4), 327–332 (2009)
44. V. Lychagin, V. Yumaguzhin, Minkowski metrics on solutions of the Khokhlov-Zabolotskaya equation. Lobachevskii J. Math. **30**(4), 333–336 (2009)
45. E. Schneider, Solutions of second-order PDEs with first-order quotients. Lobachevskii J. Math. **41**(12), 2491–2509 (2020)
46. R. Mrugala, Geometrical formulation of equilibrium phenomenological thermodynamics. Rep. Math. Phys. **14**(3), 419–427 (1978)
47. G. Ruppeiner, Riemannian geometry in thermodynamic fluctuation theory. Rev. Mod. Phys. **67**(3), 605–659 (1995)
48. A. Duyunova, V. Lychagin, S. Tychkov, Continuum mechanics of media with inner structures. Differ. Geom. Appl. **74**, 101703 (2021)
49. A. Duyunova, V. Lychagin, S. Tychkov, Quotient of the Euler system on one class of curves. J. Geom. Phys. **173**, 104432 (2022)
50. V. Lychagin, O. Lychagina, Finite dimensional dynamics for evolutionary equations. Nonlinear Dyn. **48**, 29–48 (2007)
51. V. Lychagin, M. Roop, Schrödinger equations on elliptic curves: symmetries, solutions and eigenvalue problem. Anal. Math. Phys. **10**, 34 (2020)
52. I. Anderson, C.G. Torre, The differential geometry package. Downloads. Paper 4 (2016). http://digitalcommons.usu.edu/dg_downloads/4

On Geometry of Mechanical Systems with Smooth and Non-smooth Nonholonomic Constraints

Maria Ulan

Abstract Mechanical systems with nonholonomic constraints can be analyzed within a comprehensive geometric framework based on Hamiltonian mechanics. The dynamics of such systems are determined by constructing vector fields on constraint submanifolds, with particular emphasis on the underlying symplectic geometry. We extend this framework to accommodate non-smooth constraints, enabling the analysis of systems exhibiting discontinuities such as collisions and sliding phenomena. Two mechanical systems are studied in detail: a pendulum with a sliding mass, and an inverted pendulum with an oscillating suspension point. For the sliding mass pendulum, we establish stability properties of the vertical position and identify specific nonholonomic constraints that achieve asymptotic stabilization. For the inverted pendulum, we derive the Mathieu equation directly from the geometric formulation and recover the classical stability criterion that accounts for the counterintuitive stabilization of the upright position via rapid oscillations. This methodology provides a unified treatment of both smooth and non-smooth dynamics, offering insights into control strategies for mechanical systems and illustrating the power of geometric methods in analyzing complex dynamical behavior.

1 Introduction

From a mathematical point of view, to study the motion of a given mechanical system means to study the Hamiltonian system on $T^*(M)$, whose Hamiltonian function is the total mechanical energy of the considered system, and M is its configuration space. This approach to classical mechanics can be called Hamiltonian, or Hamiltonian mechanics.

M. Ulan (✉)
Baltic Institute of Mathematics, Warsaw, Poland
e-mail: maria.ulan@baltinmat.eu

M. Ulan and N. C. Combe (eds.), *Geometric Methods in Physical Systems: From Differentiable Structures to Applications*, Tutorials, Schools, and Workshops in the Mathematical Sciences, https://doi.org/10.1007/978-3-032-00399-7_4

One example of this kind is geometric optics. The equation describing the propagation of light rays in an inhomogeneous and anisotropic medium has the form:

$$\sum_{i=1}^{3} \frac{1}{\alpha_i(x)} \left(\frac{\partial f}{\partial x_i} \right)^2 = 1, \tag{1}$$

where $\alpha_i(x)$ is the speed of light propagation at point x along the Ox_i axis.

Equation (1) is the Hamilton-Jacobi equation, corresponding to the Hamiltonian function $H = \frac{p_1^2}{\alpha_1} + \frac{p_2^2}{\alpha_2} + \frac{p_3^2}{\alpha_3}$. The corresponding Hamiltonian system is a system of mechanical type, due to which an optico-mechanical analogy arises, playing an exceptionally important role in the development of a whole range of areas of physics and mathematics. In fact, it was this analogy that led to the creation of Hamiltonian formalism [1]. It also played a leading role ind discovery of thr famous equation of non-relativistic quantum mechanics, e.g., the Schrödinger equation [2].

Here's another example. The symbol of a linear differential operator D on a manifold M can be understood as a function on $T^*(M)$, i.e., as a Hamiltonian (for more details see [3]). Formally, this is achieved by replacing the highest-order derivatives in the operator D with $\frac{\partial}{\partial x_i} \to p_i$. The properties of the resulting Hamiltonian system that arises in this way are closely related to the properties of the local solvability of the operator D [4].

The problem of describing the motion of a subsystem often arises in various mechanisms. Such mechanisms are frequently systems with nonholonomic constraints imposed on them. Despite the fact that such systems have been studied for quite some time, a general theory describing their motion has not yet been developed. Furthermore, there are many substantive questions related to the study of these systems. For describing the motion of a system with nonholonomic constraints, it is convenient to use the Hamiltonian formalism. In this case, the formalism itself should be understood from the perspective of the theory of differential operators. This approach allows to obtain a natural and clear description of the system (for more details, see [5]). With this approach, geometric concepts and considerations play an important role in the description of the system. In particular, significant attention is paid to symplectic geometry and the geometry of dynamical systems.

Motion of a mechanical system with nonholonomic constraints can be described in the language of Hamiltonian mechanics; the general algorithm can be found in [4, 6]. This algorithm provides broad possibilities for describing such systems. In this work, we start with setting the scene and discussing the general construction in Sect. 2. Then in Sect. 3 we apply this algorithm to the motion of a pendulum with a sliding mass and in Sect. 4 to the inverted pendulum. The phase space and the vector field defining the trajectories of motion on it are constructed for these systems. As a result of the analysis of the vector field, it was possible to draw conclusions regarding the stability of these systems in the atypical vertical position for a pendulum. A method for constructing the vector field was also developed and used in some cases when the nonholonomic constraint is given by a non-smooth function. This was achieved due to the connection between the vector field and the trajectories along it. Discontinuous nonholonomic constraints in a mechanical system can be observed during collisions between parts of the system, see [7]. Non-smooth constraints can occur when the system is in a sliding control mode. The study of such systems is closely related to the theory of optimal control, in which they play an important role.

2 Geometry of Nonholonomic Mechanical Systems

In this section, we present a Hamiltonian framework for analyzing mechanical systems subject to nonholonomic constraints. This approach employs the tools of symplectic geometry to formulate the dynamics of constrained mechanical systems in a coordinate-invariant manner. We begin by establishing the fundamental concepts of configuration space, phase space, and generalized Hamiltonian. Building on this foundation, we then present a systematic algorithm for constructing appropriate vector fields that describe the dynamics of systems with smooth nonholonomic constraints. Finally, we extend our analysis to handle the more challenging case of non-smooth constraints, which are essential for modeling mechanical systems involving collisions and sliding phenomena. This Hamiltonian perspective provides a unified treatment of various mechanical systems and offers powerful analytical tools for understanding their behavior.

2.1 Basic Concepts

We begin by establishing the fundamental mathematical framework for describing mechanical systems. This section introduces the key geometric structures that will enable us to formulate and analyze the dynamics of constrained mechanical systems.

Let's denote by M a manifold equipped with a metric, which we will call the *configuration space*. The metric defines a kinetic energy function G on M. The *phase space*, denoted by Φ, is a manifold isomorphic to the cotangent bundle of M, capturing both position and momentum information.

Given a local chart (U, x) on M, we can construct the canonical chart $(T^*U, (x, p))$ on Φ, where p represents the momentum variables. These momentum variables are determined by the Legendre transformation L according to

$$p_i = \frac{\partial G}{\partial \dot{x}_i}.$$

Here, the coordinates are expressed as

$$x = (x_1, \ldots, x_n) \quad \text{and} \quad p = (p_1, \ldots, p_n),$$

where n is the dimension of the configuration space M.

To facilitate our analysis, we introduce two important mappings in the chart. The natural projection π maps points in phase space to their corresponding configurations:

$$\pi : \Phi \to M, \quad (x, p) \mapsto x.$$

Additionally, we define the natural correspondence I that associates differential forms with vector fields:

$$I : \Lambda(\Phi) \to D(\Phi)$$

$$\sum_{i=1}^{n} a_i(x, p)dx_i + b_i(x, p)dp_i \mapsto \sum_{i=1}^{n} a_i(x, p)\frac{\partial}{\partial x_i} + b_i(x, p)\frac{\partial}{\partial p_i}.$$

Central to our formulation is the canonical 2-form $\Omega = \sum_{i=1}^{n} dp_i \wedge dx_i$, which endows Φ with a symplectic structure essential for Hamiltonian mechanics.

We define the function $\tilde{G}$ on Φ to satisfy the relation $G = \tilde{G} \circ L$, establishing a correspondence between energy representations in different spaces.

The 1-form E on Φ is called the *generalized Hamiltonian* if its components satisfy $E_{n+i} = \frac{\partial \tilde{G}}{\partial p_i}$ in the chosen chart.

With these structures in place, we can now formally define a mechanical system as a symplectic manifold Φ equipped with a 2-form Ω and a covector field E. Each point in Φ represents a complete state of the system, specifying both its configuration and momentum. The forms Ω and E together determine the trajectories on Φ that describe the dynamics of the system.

When constraints are present, we describe the motion of a constrained subsystem by restricting the system to states corresponding to points on a submanifold:

$$\Pi = \{b \in \Phi \mid F(x(b), p(b)) = 0\}, \tag{1}$$

where $F(x, p)$ is a piecewise continuous function in x and piecewise smooth in p. We refer to $F(x, p)$ as a *nonholonomic constraint*. The discontinuities in F correspond to collisions in the system at specific points, while the non-smoothness in F models sliding behaviors.

To fully characterize the dynamics of this constrained system, we must construct a vector field over Π derived from the generalized Hamiltonian E. For systems with smooth control constraints, this construction follows the algorithm described in [7].

2.2 Construction of Vector Fields for Systems with Nonholonomic Constraints

The algorithm for constructing appropriate vector fields on constrained manifolds has been extensively described in [4, 6]. In this subsection, we present a systematic approach to constructing these vector fields, which are essential for describing the dynamics of constrained mechanical systems.

Consider a symplectic space (Φ, Ω) equipped with a generalized Hamiltonian E. We focus on the case where the nonholonomic constraint can be expressed in an affine form with respect to $\frac{\partial \tilde{G}}{\partial p_i}$, with coefficients that are independent of the momentum variables p:

$$F(x, p) = a(x)\frac{\partial \tilde{G}}{\partial p_i} + b(x).$$

This constraint defines a smooth submanifold $\Pi \subset \Phi$ characterized by Eq. (1). Our goal is to construct a vector field on Π that describes the dynamics of the constrained system. To achieve this, we proceed as follows.

First, at each point $b \in \Pi$, we identify the tangent subspace $T_b\Pi$, which consists of all tangent vectors satisfying the differential of the constraint:

$$T_b\Pi = \{\xi \in T_b\Phi \mid dF|_b(\xi) = 0\}.$$

From the inverse Lagrange transformation, we obtain a relationship between velocity and momentum variables:

$$\dot{x}_i = dx_i^{\Phi}(\dot{x}, \dot{p}) = dx_i^{M} \circ d\pi(\dot{x}, \dot{p}) = \frac{\partial \tilde{G}}{\partial p_i}.$$

This relationship allows us to define a specialized distribution within $T_b\Pi$ that corresponds to the imposed constraints. We denote this distribution as $T_b^0\Pi \subset T_b\Pi$:

$$T_b^0\Pi = \{\xi \in T_b\Pi \mid a(x)\dot{\xi}_i + b_i = 0\}.$$

It is important to note that $T_b^0\Pi$ forms an affine subspace within the linear space $T_b\Pi$. To facilitate our analysis, we construct the corresponding linear subspace $T_b^1 = \rho + T_b^0$, where $\rho \in T_b\Pi$ represents a section of the subspace such that:

$$T_b^1\Pi = \{\xi \in T_b\Pi \mid dF|_b(\xi) = 0, \sigma(\xi) = 0\}.$$

Here, $\sigma = a(x)dx$ is a 1-form on Φ derived from the coefficients of the constraint equation.

Since both dF and σ exhibit degeneracy on $T^1\Pi$, we introduce a factorization procedure. Let $\omega \in \Lambda(\Phi)/\{dF = 0; \sigma = 0\}$ represent the equivalence class of E under this factorization.

We consider the 2-form Ω to be non-degenerate on $T_b^0\Pi$ if it maintains non-degeneracy when restricted to $T_b^1\Pi$. Specifically, this non-degeneracy condition is met when:

$$\mathrm{Ker}\Omega|_{T_b^1\Pi} = \{\eta \in T_b^1\Pi \mid (i_\eta\Omega)|_{T_b^1\Pi} = 0\} = 0.$$

When Ω satisfies this non-degeneracy condition on $T_b^0\Pi$, we can uniquely define a covector $E_b^0 \in \omega$ that generates a vector:

$$X_b^0 = I|_b(i_{I|_b(E_b^0)}\Omega|_{T_b\Phi}).$$

Consequently, if Ω is non-degenerate on $T_b^0\Pi$ for all points $b \in \Pi$, we can define a complete vector field X^0 across the entire submanifold Π. This vector field captures the dynamics of the mechanical system subjected to the given nonholonomic constraints.

Remark 4.1 In cases where the 2-form Ω becomes degenerate on $T_b^0\Pi$ at certain points $b \in \Pi$, the vector field X^0 can still be extended by continuity, allowing us to maintain a description of the system dynamics throughout the submanifold.

2.3 *Extension to Non-smooth Nonholonomic Constraints*

In this section, we extend our analysis to consider piecewise continuous and piecewise-smooth surfaces in the phase space that define nonholonomic constraints of a mechanical system. This extension is particularly important for modeling discontinuous phenomena such as sliding and collision mechanics.

Let us consider two surfaces corresponding to different nonholonomic constraints with a common boundary:

$$\Pi^> = \{(x_1, \ldots, x_n, p_1, \ldots, p_n) \mid \frac{\partial \tilde{G}}{\partial p_j} = f(x_1, \ldots, x_n); (x_1, \ldots, x_n) \in M^> \subset M\}$$

$$\Pi^< = \{(x_1, \ldots, x_n, p_1, \ldots, p_n) \mid \frac{\partial \tilde{G}}{\partial p_j} = g(x_1, \ldots, x_n); (x_1, \ldots, x_n) \in M^< \subset M\}.$$

We assume the following conditions for the boundary between these constraint surfaces:

- Let the point $y \in \partial M^> \cap \partial M^<$ and there exists a neighborhood $U \ni y$ such that

$$U \cap \partial M^> = U \cap \partial M^<.$$

- There exists a smooth function P on M defining the submanifold Y:

$$Y = \{(x_1, \ldots, x_n) \in U \mid P(x_1, \ldots, x_n) = 0\}.$$

- For the piecewise-smooth manifold Π, the constraint functions must agree on the boundary:

$$f(x_1, \ldots, x_n) = g(x_1, \ldots, x_n) \quad \forall (x_1, \ldots, x_n) \in Y.$$

Using the algorithm from [6], we can construct vector fields $Z^{>}(\omega)$ and $Z^{<}(\omega)$ at points of the submanifolds $\Pi^{>}$ and $\Pi^{<}$, respectively. These vector fields are obtained by projecting the unconstrained vector field onto the smooth submanifolds defined by the nonholonomic constraints.

For a point on the boundary:

$$b_0 \in \Pi \cap \pi^{-1}(Y) = \Pi^{>} \cap \Pi^{<} \cap \pi^{-1}(U)$$

two vectors $Z^{>}(\omega)_{b_0}$ and $Z^{<}(\omega)_{b_0}$ correspond to it. The following proposition establishes an important relationship between these vectors.

Proposition 4.1 *Under the assumptions made above, if one of the two vectors* $Z^{>}(\omega)_{b_0}$ *or* $Z^{<}(\omega)_{b_0}$ *lies in the intersection of the tangent subspaces* $T_{b_0}\Pi^{>}$ *and* $T_{b_0}\Pi^{<}$, *then* $Z^{>}(\omega)_{b_0} = Z^{<}(\omega)_{b_0}$.

Proof Let us define the constraint functions:

$$F(x_1, \ldots, x_n, p_1, \ldots, p_n) = \frac{\partial \tilde{G}}{\partial p_j} - f(x_1, \ldots, x_n)$$

$$G(x_1, \ldots, x_n, p_1, \ldots, p_n) = \frac{\partial \tilde{G}}{\partial p_j} - g(x_1, \ldots, x_n).$$

The tangent spaces at b_0 are characterized by

$$T_{b_0}\Pi^{>} = \{\xi \in T_{b_0}\Phi \mid dF|_{b_0}(\xi) = 0\}$$

$$T_{b_0}\Pi^{<} = \{\xi \in T_{b_0}\Phi \mid dG|_{b_0}(\xi) = 0\}.$$

Then

$$T_{b_0}\Pi^{>} \cap T_{b_0}\Pi^{<} = \{\xi \in T_{b_0}\Phi \mid dF|_{b_0}(\xi) = 0, dG|_{b_0}(\xi) = 0\}.$$

Let $X_{b_0} = I(i_{I(E|_{b_0})}\Omega)$ be the vector corresponding to the Hamiltonian in the unconstrained system. The constrained vector field at b_0 on $\Pi^{>}$ takes the form:

$$Z^{>}(\omega)_{b_0} = X_{b_0} + \lambda^{>}\frac{\partial}{\partial p_j},$$

where $\lambda^{>}$ is chosen such that $dF|_{b_0}(Z^{>}(\omega)_{b_0}) = 0$.

If $Z^{>}(\omega)_{b_0} \in T_{b_0}\Pi^{>} \cap T_{b_0}\Pi^{<}$, then we also have $dG|_{b_0}(Z^{>}(\omega)_{b_0}) = 0$.

Similarly, for $\Pi^{<}$, we have

$$Z^{<}(\omega)_{b_0} = X_{b_0} + \lambda^{<}\frac{\partial}{\partial p_j}.$$

Since $\lambda^< \in \mathbb{R}$ is uniquely defined by the condition $dG|_{b_0}(Z^<(\omega)_{b_0}) = 0$, and this condition already holds for the vector $Z^>(\omega)_{b_0}$, we must have $\lambda^> = \lambda^<$. Therefore, as a special case, if the tangent spaces coincide at b_0 (i.e., $T_{b_0}\Pi^> = T_{b_0}\Pi^<$), then $Z^>(\omega)_{b_0} = Z^<(\omega)_{b_0}$.

Now, let us consider the more interesting case where $T_{b_0}\Pi^> \neq T_{b_0}\Pi^<$ and neither vector lies in the intersection of the tangent subspaces, i.e., $Z^>(\omega)_{b_0}, Z^<(\omega)_{b_0} \notin T_{b_0}\Pi^> \cap T_{b_0}\Pi^<$, which implies $Z^>(\omega)_{b_0} \neq Z^<(\omega)_{b_0}$.

We can construct curves $\gamma^>(t) \subset \Pi^>$ and $\gamma^<(t) \subset \Pi^<$ along the vector fields $Z^>(\omega)$ and $Z^<(\omega)$, passing through the point b_0. Let us assume that $\gamma^>(t)$ arrives at b_0, while $\gamma^<(t)$ departs from b_0.

Since $T_{b_0}\Pi^> \neq T_{b_0}\Pi^<$, we have

$$Z^>(\omega)_{b_0}, Z^<(\omega)_{b_0} \notin T_{b_0}\Pi^> \cap T_{b_0}\Pi^< = T_{b_0}(\Pi^> \cap \Pi^<).$$

This means that trajectory $\gamma^>(t)$ ends at b_0 on $\Pi^>$, and $\gamma^<(t)$ begins at b_0 on $\Pi^<$. For a correct description of the system's dynamics at this transition point, we should choose the tangent vector $Z^<(\omega)_{b_0}$ to describe the motion after reaching the boundary.

If we can uniquely construct a tangent vector for each point of the manifold Π in this manner, then we obtain a complete description of the dynamics of the mechanical system subject to piecewise-smooth nonholonomic constraints. Such constraints naturally arise in sliding systems, where the system transitions between different constraint conditions.

Note that we do not consider cases where both $\gamma^>(t)$ and $\gamma^<(t)$ either begin or end at b_0, as these situations have unclear mechanical interpretations and are not of interest in the present work.

In the case of collisions on the mechanism or collisions between its parts, we can model the resulting dynamics using piecewise-smooth surfaces over M. An algorithm for calculating such collisions is described in [7].

3 Pendulum with Sliding Mass

The pendulum with sliding mass represents a fundamental mechanical system that exhibits complex nonholonomic dynamics and serves as an important benchmark in the study of variable-geometry mechanisms. This system consists of a pendulum with a mass that can move along the rod, creating a coupling between rotational and translational motions that leads to rich dynamical behavior and control challenges [18, 19].

In practical engineering, systems with variable-length components appear in numerous contexts, including crane operations, deployable space structures, tethered satellite systems, and certain robotic manipulators [20]. The dynamics of cable-suspended loads, which share mathematical similarities with the sliding mass pendulum, are particularly relevant in construction, shipping, and materials handling industries where stability and oscillation control remain critical challenges [21].

The sliding mass pendulum also presents a fascinating case study in nonholonomic mechanics, as the imposition of constraints on the sliding mass creates a system whose behavior cannot be fully described by position coordinates alone [22, 23]. This property makes it an ideal candidate for exploring the geometric approach to mechanics that we have developed in previous sections.

From a historical perspective, sliding mass pendulum also known as variable-length pendulum systems have been studied since the early twentieth century, with significant developments in both theoretical understanding and practical applications [19, 23]. More recently, researchers have connected these systems to modern control theory, demonstrating how nonholonomic constraints can be leveraged to achieve stability in otherwise unstable configurations [18].

In this section, we apply the geometric framework discussed in the previous section to a specific mechanical system: a pendulum with a sliding mass. This system represents an important class of nonholonomic mechanical problems and serves as an illustrative example for the theoretical approach. The sliding mass pendulum exhibits rich dynamical behavior, combining elements of oscillatory and translational motion in a coupled system that can be analyzed through the lens of symplectic geometry and Hamiltonian mechanics.

We begin by examining the unconstrained dynamics of the system, establishing the foundational mathematical description in terms of configuration space, phase space, and the Hamiltonian formulation. We then introduce various types of constraints; starting with smooth nonholonomic constraints, followed by directional constraints that distinguish between positive and negative motion along the rod, and eventually considering discontinuous scenarios involving collisions with obstacles. Through this analysis, we demonstrate how the geometric approach provides a unified treatment of both smooth and non-smooth dynamical transitions.

Of particular interest is the stability analysis of the system, where we investigate the conditions under which the vertical equilibrium position exhibits stability or instability, and explore methods for controlling the system's behavior through appropriately designed constraints. The rich dynamics of the sliding mass pendulum, including its potential for chaotic behavior, parametric resonance, and energy exchange between rotational and translational modes, continues to attract attention in theoretical mechanics. The results obtained not only validate the theoretical framework but also provide insights into the control and stabilization of nonholonomic mechanical systems.

3.1 Pendulum Without Imposed Constraints

We consider a mechanical system consisting of a weightless rod rotating in a vertical plane around a fixed pivot point, with a mass m that can slide freely along the rod. The entire system is subject to a uniform gravitational field. This configuration represents

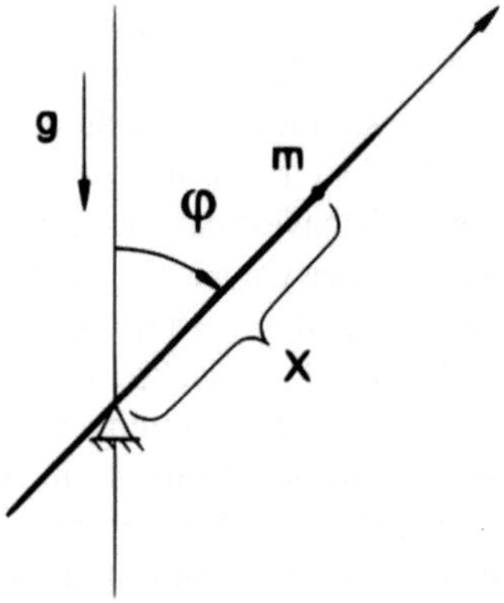

Fig. 1 Pendulum with sliding mass

a simple yet rich example of a system with coupled rotational and translational motion, see Fig. 1.

The configuration of this system is completely specified by two variables: the angle ϕ that measures the rod's inclination from the vertical axis, and the distance x of the sliding mass from the suspension point along the rod. These variables constitute our generalized coordinates.

The configuration space of this system forms a cylinder:

$$M = \{(x, \phi) \mid x \in \mathbb{R}, \phi \in S^1\}.$$

For practical analysis, we work within the chart $U = \{b \mid \phi(b) \in (-\pi, \pi)\}$, which avoids the coordinate singularity at $\phi = \pm\pi$.

The kinetic energy of the system defines a metric on the configuration space:

$$G : TM \times TM \to \mathbb{R}.$$

This metric acts according to the rule:

$$G|_{(x,\phi)}(\dot{x}, \dot{\phi}) = \frac{m\dot{x}^2}{2} + \frac{(mx^2 + J)\dot{\phi}^2}{2}.$$

Here, the first term represents the kinetic energy associated with the radial motion of the sliding mass along the rod, while the second term captures the rotational kinetic energy. The parameter J denotes the moment of inertia for rotation of the sliding mass around an axis passing through its center of mass.

For instance, if the sliding mass is modeled as a uniform rod of length l, then its moment of inertia is

$$J = \frac{ml^2}{12}.$$

In many practical scenarios where the distance from the pivot is much greater than the size of the sliding mass ($l \ll x$), we can approximate:

$$J \ll mx^2.$$

The potential energy of the system, denoted by a smooth function V on M, arises from the gravitational field and is given by

$$V(x, \phi) = mgx \cos(\phi).$$

This expression captures how the potential energy varies with both the distance of the sliding mass from the pivot and the rod's angular position. When $\phi = 0$ (rod pointing downward), the potential energy is minimized at $-mgx$, while at $\phi = \pi$ (rod pointing upward), it reaches a maximum of mgx.

To transition to the Hamiltonian formulation, we need to work in the phase space $\Phi = T^*M$, which is equipped with a natural symplectic structure (Φ, Ω). The phase space describes not only the configuration of the system but also its momentum state.

We use the Legendre transformation associated with the metric G to obtain the canonical coordinates of the phase space:

$$L : TM \to T^*M$$

$$(x, \phi, \dot{x}, \dot{\phi}) \mapsto (x, \phi, p, \iota).$$

Here, p and ι represent the generalized momenta corresponding to configuration variables:

- $p = \frac{\partial G}{\partial \dot{x}} = m\dot{x}$ is the linear momentum of the sliding mass along the rod.
- $\iota = \frac{\partial G}{\partial \dot{\phi}} = (mx^2 + J)\dot{\phi}$ is the angular momentum of the system with respect to the suspension point.

In these canonical coordinates, the symplectic two-form takes the standard skew-diagonal representation:

$$\Omega = dp \wedge dx + d\iota \wedge d\phi.$$

The kinetic energy can be expressed as a quadratic function $\tilde{G}$ on the phase space:

$$\tilde{G} = \frac{p^2}{2m} + \frac{\iota^2}{2(mx^2 + J)}.$$

This form highlights how the kinetic energy depends on both the linear momentum p and the angular momentum ι, with the second term showing the coupling between rotational motion and position of the sliding mass.

The potential energy function is simply lifted from the configuration space M to the phase space Φ via the natural projection.

Combining the kinetic and potential energies, we obtain the Hamiltonian function for the system:

$$H = \tilde{G} + V \circ \pi = \frac{p^2}{2m} + \frac{\iota^2}{2(mx^2 + J)} + mgx\cos(\phi).$$

This Hamiltonian fully characterizes the dynamics of the unconstrained pendulum with a sliding mass.

Finally, we construct the generalized Hamiltonian one-form E corresponding to the function H:

$$E = \frac{\partial \tilde{G}}{\partial p}dp + \frac{\partial \tilde{G}}{\partial \iota}d\iota + \left(\frac{\partial \tilde{V}}{\partial x} + \frac{\partial \tilde{G}}{\partial x}\right)dx + \frac{\partial \tilde{V}}{\partial \phi}d\phi =$$

$$= \frac{p}{m}dp + \frac{\iota}{mx^2 + J}d\iota + \left(mg\cos(\phi) - \frac{mx\iota^2}{(mx^2 + J)^2}\right)dx - mgx\sin(\phi)d\phi.$$

This one-form encodes the complete dynamics of the system in a coordinate-invariant manner. The equations of motion can be derived from it using the framework developed in the previous section.

The coefficients of dp and $d\iota$ represent the velocities $\dot{x}$ and $\dot{\phi}$, respectively, while the coefficients of dx and $d\phi$ capture the forces and torques acting on the system. Particularly noteworthy is the term $-\frac{mx\iota^2}{(mx^2+J)^2}$ in the coefficient of dx, which represents the centrifugal force acting on the sliding mass due to the rotation of the rod.

3.2 Smooth Nonholonomic Constraint

To study the dynamics of a pendulum with a sliding mass under constraints, we must formulate and analyze the problem within the geometric framework established earlier. Let us consider a smooth submanifold in the phase space:

$$\Pi = \{(x, \phi, p, \iota) \in \Phi \mid F(x, \phi, p, \iota) = 0\}.$$

Each point on this submanifold represents a possible state of a mechanical system subject to a nonholonomic constraint. For analysis, we focus on a constraint of the form $F(x, \phi, p, \iota) = p - f(x)$, where $f \in C^\infty(\mathbb{R})$. This constraint links the momentum p directly to a function of position, effectively restricting the system's motion.

To characterize the system's behavior over time, we need to construct a trajectory along a vector field that is both consistent with the Hamiltonian H and confined to the submanifold Π. We begin by constructing the tangent bundle on Π:

$$T\Pi = \coprod_{b \in \Pi} T_b\Pi = \{(b, \xi) \mid b \in \Pi, dF|_b(\xi) = 0\},$$

where $dF = dp - f'(x)dx$ is a covector field on Φ.

Among all tangent vectors to the submanifold Π, we are particularly interested in those that properly respect the nonholonomic constraint under the Legendre transformation. These vectors form a special distribution $T^0\Pi$:

$$T^0\Pi = \{\varsigma \in T \mid F \circ L \circ d\pi(\varsigma) = 0\}$$
$$T^0\Pi = \coprod_{b \in \Pi} T_b^0\Pi.$$

At each point $b \in \Pi$, this distribution is characterized by

$$\begin{aligned} T_b^0\Pi &= \{\xi \in T_b\Pi \mid F|_{\pi(b)} \circ L \circ d\pi|_{\pi(b)}(\xi) = 0\} = \\ &= \{\xi \in T_b\Phi \mid dF|_b(\xi) = 0; F|_{\pi(b)} \circ L \circ d\pi|_{\pi(b)}(\xi) = 0\} = \\ &= \{\xi \in T_b\Phi \mid dF|_b(\xi) = 0; \sigma|_b(\xi) - f(b_1) = 0\} = \\ &= \{\xi \in T_b\Phi \mid \xi_3 - f'(b_1)\xi_1 = 0; m\xi_1 - f(b_1) = 0\}, \end{aligned}$$

where $b = (x, \phi, p, \iota) \in \Pi, \xi = (\dot{x}, \dot{\phi}, \dot{p}, \dot{\iota}) \in T_b\Phi$, and $\sigma = mdx$ is a covector field on Φ.

The covectors $dF|_b$ and $\sigma|_b$ are linearly independent for any point $b \in \Pi$, which means the resulting distribution $T^0\Pi$ has a codimension of two. Geometrically, $T_b^0\Pi$ forms an even-dimensional affine subspace of the vector space $T_b\Phi$.

To facilitate analysis, we construct a corresponding linear subspace $T_b^1\Pi$ by shifting the affine subspace $T_b^0\Pi$ by a vector $r \in T_b\Phi$:

$$T_b^1\Pi = T_b^0\Pi + r = \{\xi \in T_b\Phi \mid dF|_b(\xi) = 0; \sigma|_b(\xi) = 0\}.$$

Note that while the shift vector r is not uniquely defined, the resulting linear subspace is well defined.

We now consider the space dual to $T_b^1\Pi$:

$$T_b^{1^*}\Pi = T_b^*\Phi/\{dF = 0, \sigma = 0\}.$$

To determine the dynamics on the constraint manifold, we need to narrow down the generalized Hamiltonian to $T_b^1\Pi$. This produces an equivalence class of one-forms:

$$\omega|_b = \overline{E|_b} = \{E|_b + \mu(b)dF|_b + \lambda(b)\sigma|_b \mid \mu(b), \lambda(b) \in \mathbb{R}\}.$$

That is

$$\omega = \overline{\{E + \mu dF + \lambda\sigma\}},$$

where μ and λ are functions on Π.

To ensure that our construction yields a well-defined vector field, we must verify that the symplectic form Ω remains non-degenerate when restricted to $T_b^1\Pi$. We examine the kernel:

$$\begin{aligned}\mathrm{Ker}(\Omega|_{T_b^1\Pi}) &= \{\eta \in T_b^1\Pi \mid i_\eta \Omega|_{T_b^1\Pi} = 0\} = \\ &= \{\eta \in T_b^1\Pi \mid \Omega(\eta, \xi) = 0, \xi \in T_b^1\Pi\}.\end{aligned}$$

For any $\eta \in T_b^1\Pi$, the condition becomes

$$\begin{aligned}\Omega(\eta, \xi) &= \eta_3\xi_1 + \eta_4\xi_2 - \eta_1\xi_3 - \eta_2\xi_4 = \\ &= \eta_4\xi_2 - \eta_2\xi_4 = 0 \quad \forall \xi \in T_b^1\Pi.\end{aligned}$$

This condition is satisfied if and only if $\eta = 0$, which means $\mathrm{Ker}(\Omega|_{T_b^1\Pi}) = \{0\}$. Therefore, Ω is indeed non-degenerate on $T_b^1\Pi$.

This non-degeneracy allows us to uniquely construct a vector $Z(\omega)|_b \in T_b^0\Pi$ corresponding to the covector $\omega|_b$:

$$\begin{aligned}Z(\omega) &= I(i_{I(E+\mu dF+\lambda\sigma)}\Omega) = \\ &= \left(\frac{p}{m} - \mu, \frac{\iota}{mx^2 + J}, \frac{mx\iota^2}{(mx^2 + J)^2} - mg\cos(\phi) - \mu f'(x) - \lambda m, mgx\sin(\phi)\right).\end{aligned}$$

For $Z(\omega)$ to lie in $T^0\Pi$, it must satisfy certain conditions. These conditions, combined with the constraint equation $p - f(x) = 0$, yield

$$\begin{aligned}0 &= \sigma(Z(\omega)) - f(x) = m\left(\frac{p}{m} - \mu\right) - f(x) \\ 0 &= dF(Z(\omega)) = \frac{mx\iota^2}{(mx^2 + J)^2} - mg\cos(\phi) - \mu f'(x) - \lambda m - f'(x)\left(\frac{p}{m} - \mu f'(x)\right).\end{aligned}$$

Solving these equations for the Lagrange multipliers μ and λ:

$$\begin{aligned}\mu &= 0 \\ \lambda &= \frac{x\iota^2}{m(mx^2 + J)^2} - g\cos(\phi) - f'(x)\frac{p}{m^2}.\end{aligned}$$

Substituting these values back into our expression for $Z(\omega)$, we obtain the uniquely determined vector field on Π:

$$Z(\omega) = \left(\frac{p}{m}, \frac{\iota}{mx^2 + J}, f'(x)\frac{p}{m}, mgx\sin(\phi)\right).$$

This vector field completely characterizes the dynamics of the pendulum with a sliding mass subject to the nonholonomic constraint $p = f(x)$. The components of this vector field represent the rates of change of position, angle, linear momentum, and angular momentum, respectively.

3.3 Directional Constraints

Having established the framework for handling single constraints, we now introduce a more sophisticated control mechanism that distinguishes between different directions of motion along the rod. This approach allows us to create oscillatory behaviors of the sliding mass, which have important applications in various mechanical systems and control problems.

We define different constraints on the momentum coordinate p depending on its sign:

$$\begin{aligned} p - f(x) = 0 \quad &\text{if } p \geq 0 \\ p - g(x) = 0 \quad &\text{if } p \leq 0, \end{aligned}$$

where $f(x) \geq 0$ and $g(x) \leq 0$ are piecewise-smooth functions on $\mathbb{R}$.

These direction-dependent constraints have clear physical interpretations. The first equation applies when the sliding mass moves in the positive x direction along the rod (away from the pivot), while the second equation applies when the mass moves in the negative direction (toward the pivot). By carefully designing the functions $f(x)$ and $g(x)$, we can create predictable oscillatory patterns or other desired behaviors.

Each constraint equation defines a distinct surface in its corresponding half-space of the phase space, which we denote as Π^+ and Π^- respectively. Together, they form a piecewise-smooth submanifold $\Pi = \Pi^+ \cup \Pi^-$ that represents all possible states of the controlled system.

The algorithm we developed in the previous subsection for constructing vector fields on constraint manifolds depends only on local properties: the local geometry of the submanifold Π, the coordinates of the point under consideration, and the value of the generalized Hamiltonian at that point. In the present context, this locality property offers a distinct advantage.

For any point b that lies strictly within either Π^+ or Π^- (i.e., not on the boundary between them), we can directly apply the previously developed algorithm, considering only the relevant constraint. This yields a well-defined vector field $Z_b(\omega)$ at such points.

The critical aspect of this analysis lies in understanding what happens at the transition between the two surfaces, that is, when the sliding mass changes direction of motion along the rod. At these transition points, where $p = 0$ and simultaneously $f(x) = 0$ and $g(x) = 0$ must hold, the dynamics require special consideration to ensure continuity of the physical motion.

This piecewise approach to constraints enables us to model systems with asymmetric behaviors depending on the direction of motion, such as those with different friction coefficients for forward and backward movement, or mechanisms that intentionally resist motion in one direction more than the other. The resulting oscillatory patterns can be precisely controlled by appropriate selection of the constraint functions $f(x)$ and $g(x)$.

3.4 Collision of the Slider with an Obstacle

We now extend the analysis to consider the case where the sliding mass collides with a physical obstacle positioned along the rod. This scenario introduces discontinuities in the system's evolution and requires special mathematical treatment. The collision process creates an instantaneous change in momentum while preserving certain physical properties such as energy.

Let us consider an obstacle that limits the position of the sliding mass along the rod from above. The set of allowable states of the system is restricted to a subset of the phase space:

$$\Phi' = \{(x, \phi, p, \iota) \in \Phi \mid x \leq x_0\},$$

where x_0 represents the coordinate of the obstacle along the rod.

The boundary of this restricted phase space corresponds precisely to states where the sliding mass is in contact with the obstacle:

$$\Gamma = \partial\Phi' = \{(x, \phi, p, \iota) \in \Phi \mid x = x_0\}.$$

To analyze the dynamics at collision, we construct the tangent bundle over this boundary:

$$T\Gamma = \coprod_{b \in \Gamma} T_b\Gamma.$$

At each point $b \in \Gamma$, the tangent space is

$$T_b\Gamma = \{\xi \in T_b\Phi \mid \xi_1 = 0\}.$$

This reflects the geometric constraint that at the moment of collision, motion in the direction perpendicular to the obstacle is instantaneously restricted. Note that $T_b\Gamma$ forms a linear subspace of $T_b\Phi$ with codimension one.

A critical observation is that the canonical symplectic form Ω becomes degenerate when restricted to Γ. To characterize this degeneracy, we examine the kernel of this restriction:

η satisfies

$$\Omega_b(\eta, \xi) = \eta_3\xi_1 + \eta_4\xi_2 - \eta_1\xi_3 - \eta_2\xi_4 = 0 \quad \forall \xi \in T_b\Gamma$$

if and only if

$$\eta \in K_b = \{(0, 0, k, 0) \mid k \in \mathbb{R}\}.$$

Since $K_b \subset T_b\Gamma$, we have

$$\mathrm{Ker}(\Omega|_{T_b\Gamma}) = \{\eta \in T_b\Gamma \mid i_\eta \Omega|_{T_b\Gamma} = 0\} = \{\eta \in T_b\Gamma \mid \Omega_b(\eta, \xi) = 0 \quad \forall \xi \in T_b\Gamma\} = K_b.$$

Using this kernel, we can construct a vector field on Γ consisting of vectors from $\mathrm{Ker}(\Omega|_{T_b\Gamma})$:

$$N_b = (0, 0, 1, 0).$$

This vector field defines curves on Γ that represent instantaneous changes in momentum during collision:

$$l_{b_0} = \{(x_0, \phi_0, p_0 + \kappa, \iota_0) \in \Gamma \mid (x_0, \phi_0, p_0, \iota_0) = b_0 \in \Gamma, \kappa \in \mathbb{R}\},$$

where b_0 is a point corresponding to collision with the obstacle.

These curves form a foliation of Γ into lines l_{b_0}. By factorizing Γ with respect to this foliation (identifying points on the same line), we can define a reduced space where two-forms and a Hamiltonian correctly describe the collision dynamics. The movement between identified points is considered instantaneous, representing the sudden change in momentum during collision.

Since states in Φ' are restricted to positions where $x < x_0$, the sliding mass can only approach the obstacle with positive velocity along the rod. Given that the generalized momentum $p = m\dot{x}$, the collision process can only occur when $p_0 \geq 0$. Similarly, the rebound at a point $b_1 = (x_1, \phi_1, p_1, \iota_1) \in \Gamma$ occurs with $p_1 \leq 0$. For an obstacle that limits motion from below, the situation would be reversed: collisions would correspond to negative momentum and rebounds to positive momentum.

Recalling the directional constraints, for non-negative p the system is confined to the submanifold Π^+, and for non-positive p it is confined to Π^-. In the collision scenario, we generally do not have $f(x_0) = 0$ and $g(x_0) = 0$ simultaneously.

Therefore, a collision point in phase space has coordinates $b_0 = (x_0, \phi_0, f(x_0), \iota_0) \in \Pi^+ \cap \Gamma$, where ϕ_0 and ι_0 can take arbitrary values. This point is associated with the rebound point $b_1 = (x_0, \phi_0, g(x_0), \iota_0) \in \Pi^- \cap \Gamma$.

An important physical consideration is that if $H(b_0) = H(b_1)$, which occurs when $f(x_0) = g(x_0)$, then the collision is perfectly elastic, conserving the system's total energy.

In summary, our model represents the collision process as an instantaneous transition from one constraint surface to another, with only the generalized momentum p changing during this transition. This approach allows us to incorporate discontinuous events like collisions into the continuous dynamical framework.

3.5 Smooth Constraint Transitions

Let's study a special case where the constraint functions smoothly transition at a specific point on the rod. Assume that both constraint functions vanish at the x_0:

$$f(x_0) = g(x_0) = 0.$$

This represents a scenario where the sliding mass can momentarily come to rest at position x_0 before potentially changing direction. If both functions are smooth, i.e., $f, g \in C^\infty[0, x_0]$, then their derivatives $f'(x_0)$ and $g'(x_0)$ are uniquely defined and finite.

For this special case, let's analyze the behavior at the intersection point $b_0 = (x_0, \phi_0, p_0, \iota_0) \in \Pi^+ \cap \Pi^-$. We begin by examining the differential forms at this point:

$$dF|_{b_0} = dp - f'(x_0)dx$$
$$\sigma^+|_{b_0} = mdx - f(x_0) = mdx.$$

The tangent space for the positive constraint direction at b_0 is

$$\begin{aligned} T^1_{b_0}\Pi^+ &= \{\xi \in T_{b_0}\Phi \mid dF|_{b_0}(\xi) = 0, \sigma^+|_{b_0}(\xi) = 0\} \\ &= \{\xi \in T_{b_0}\Phi \mid \xi_3 - f'(b_0)\xi_1 = 0, m\xi_1 = 0\} \\ &= \{\xi \in T_{b_0}\Phi \mid \xi_1 = \xi_3 = 0\}. \end{aligned}$$

For any vector $\xi \in T^1_{b_0}\Pi^+$, we can verify

$$dG|_{b_0}(\xi) = \xi_3 - g'(x_0)\xi_1 = 0.$$

This implies that $T^1_{b_0}\Pi^+ \subset T_{b_0}\Pi^-$.

Furthermore, the vector $Z\omega^+_{b_0} \in T^1_{b_0}\Pi^+ \subset T_{b_0}\Pi^+ \cap T_{b_0}\Pi^-$. Since $b_0 \in \Pi^+ \cap \Pi^-$ with $f(x_0) = g(x_0)$, and the vectors $Z\omega^+_{b_0}$ and $Z\omega^-_{b_0}$ are uniquely determined, we can apply Proposition 4.1 to conclude that the vector corresponding to the imposed constraints and the system's Hamiltonian at point b_0 is uniquely determined:

$$Z(\omega)_{b_0} = Z\omega^+_{b_0} = Z\omega^-_{b_0}.$$

Computing this vector explicitly:

$$Z(\omega)_{b_0} = \left(0, \frac{\iota}{mx_0^2 + J}, 0, mgx_0\sin(\phi)\right).$$

This result has a clear physical interpretation: at the position x_0 where both constraint functions vanish, the sliding mass comes to a complete stop along the rod ($\dot{x} = 0$), while the rod itself continues its angular motion. This represents a smooth transition between the two constraint regimes.

A singular case arises when the derivative $f'(x_0) = -\infty$. In this scenario:

$$\begin{aligned}
T_{b_0}\Pi^+ &= \{\xi \in T_{b_0}\Phi \mid dx(\xi) = 0\} \\
\sigma &= mdx \\
T^1_{b_0}\Pi^+ &= \{\xi \in T_{b_0}\Phi \mid dx(\xi) = 0, \sigma|_{b_0}(\xi) - f(x_0) = 0\} = \\
&= \{\xi \in T_{b_0}\Phi \mid dx(\xi) = 0\} = T_{b_0}\Pi^+.
\end{aligned}$$

Since $T^1_{b_0}\Pi^+$ has codimension one and the two-form becomes degenerate when restricted to this space, the vector $Z\omega^+_{b_0}$ is not uniquely determined. This singularity reflects physical situations where the transition between constraint regimes becomes discontinuous, requiring additional considerations for a complete dynamical description.

3.6 *Sliding Mass Motion Along the Rod*

Having established the geometric framework for the constrained pendulum with sliding mass, we now analyze the resulting dynamics and develop explicit solutions for the system's evolution over time. This analysis will provide insights into how the imposed constraints determine the motion patterns of the sliding mass along the rod.

Consider the vector field we constructed on the submanifold defined by a single constraint:

$$\Pi = \{(x, \phi, p, \iota) \mid p - f(x) = 0\}.$$

The equations of motion on this submanifold are

$$\begin{cases}
\dot{x} = \dfrac{p}{m} \\
\dot{\phi} = \dfrac{\iota}{mx^2 + J} \\
\dot{p} = f'(x)\dfrac{p}{m} \\
\dot{\iota} = mgx\sin(\phi).
\end{cases} \tag{2}$$

Since the constraint $p - f(x) = 0$ directly relates momentum to position, we can express the velocity of the sliding mass as

$$\dot{x} = \frac{f(x)}{m}.$$

This first-order differential equation can be solved by separation of variables:

$$m\int \frac{dx}{f(x)} = t.$$

Given the function $f(x)$ and appropriate initial conditions, the position x can be expressed as a function of time:

$$x = \chi(t + c),$$

where c is a constant determined by the initial conditions.

Now, let us consider the more complex case where the constraints distinguish between directions of motion. Recall that the system is described by two surfaces in the phase space, Π^+ and Π^-, with uniquely defined vector fields projected onto their tangent spaces.

For motion in the positive direction:

$$\Pi^+ = \{(x, \phi, p, \iota) \mid p - f(x) = 0; x \in [x_{\min}, x_{\max}]\},$$

where

$$f(x) \geq 0; \quad x \in [x_{\min}, x_{\max}].$$

The corresponding vector field on Π^+ is given by Eq. (2). The position x as a function of time can be determined by

$$m \int \frac{dx}{f(x)} = t \quad \text{for} \quad x \in [x_{\min}, x_{\max}].$$

Since $\frac{m}{f(x)} \geq 0$ for $x \in [x_{\min}, x_{\max}]$, the time function is monotonic. If the time to traverse the interval is finite, i.e., $t(x) \neq \infty$ for all $x \in [x_{\min}, x_{\max}]$, then we can define the time intervals for motion along this constraint surface.

Given x_0 as the initial position of the sliding mass, we define

$$t_{\min(f)} = \int_{x_0}^{x_{\min}} \frac{m\,dx}{f(x)}$$
$$t_{\max(f)} = \int_{x_0}^{x_{\max}} \frac{m\,dx}{f(x)}.$$

The time parameter then ranges within:

$$t \in [t_{\min(f)} + c^f, t_{\max(f)} + c^f],$$

where c^f is a constant determined by initial conditions.

Remark 4.2 If there exists an $x' \in [x_{\min}, x_{\max}]$ such that $t(x') = \infty$, then the trajectory of motion will remain confined to the surface Π^+ indefinitely.

The position on the positive constraint surface can be expressed as

$$x(t) = \chi^+(t); \quad t \in [t_{\min(f)} + c, t_{\max(f)} + c].$$

Similarly, for motion on the negative constraint surface Π^-, we have

$$x(t) = \chi^-(t); \quad t \in [t_{\max(g)} + c, t_{\min(g)} + c],$$

where

$$t_{\min(g)} = \int_{x_0}^{x_{\min}} \frac{m\,dx}{g(x)}$$
$$t_{\max(g)} = \int_{x_0}^{x_{\max}} \frac{m\,dx}{g(x)}.$$

Each transition between constraint surfaces at time t' requires determining a new constant c that defines the shift in the time interval. The complete trajectory is thus a piecewise function:

$$\begin{aligned} x(t) &= \chi^+(t) \quad \text{if} \quad t \in [t_i, t_{i+1}] \\ x(t) &= \chi^-(t) \quad \text{if} \quad t \in [t_j, t_{j+1}]. \end{aligned}$$

Given the constraints imposed on the system and the initial conditions, we can fully determine how the position x evolves with time. The angular coordinate ϕ and angular momentum ι can then be calculated by integrating their respective differential equations using $x(t)$ as an input.

This piecewise solution structure provides a complete description of the system's temporal evolution under direction-dependent constraints, enabling us to analyze complex behaviors such as oscillations and limit cycles.

3.7 Stability Analysis

Having determined the motion of the sliding mass along the rod as a function of time, we now turn our attention to the angular dynamics of the pendulum system. The goal is to analyze the stability properties of the angular motion, assuming that the position function $x(t)$ is known and satisfies $x(t) \geq 0$ for all $t \in \mathbb{R}$.

The angular dynamics are governed by the following subsystem of equations:

$$\begin{cases} \dot{\phi} = \frac{\iota}{mx^2+J} \\ \dot{\iota} = mgx\sin(\phi). \end{cases}$$

To understand the stability properties near the equilibrium point $(\phi, \iota) = (0, 0)$ (which corresponds to the pendulum hanging straight down with no angular momentum), we linearize the vector field around this point. The Jacobian matrix of the linearized system is

$$J = \begin{pmatrix} 0 & \frac{1}{mx^2+J} \\ mgx\cos(\phi)|_{\phi=0} & 0 \end{pmatrix} = \begin{pmatrix} 0 & \frac{1}{mx^2+J} \\ mgx & 0 \end{pmatrix}.$$

The stability properties are determined by the eigenvalues of this matrix, which we find by solving the characteristic equation:

$$\det(J - \lambda I) = 0 \Rightarrow \lambda^2 - \frac{mgx}{mx^2 + J} = 0.$$

This yields the eigenvalues:

$$\lambda_1 = \sqrt{\frac{mgx}{mx^2 + J}}$$
$$\lambda_2 = -\sqrt{\frac{mgx}{mx^2 + J}}.$$

Since we've established that $x(t) \geq 0$ for all $t \in \mathbb{R}$, and given that m, g, and J are all positive physical parameters, we can conclude

$$\lambda_1(t), \lambda_2(t) \in \mathbb{R} \quad \forall t \in \mathbb{R}$$
$$\lambda_1(t) \geq 0$$
$$\lambda_2(t) \leq 0.$$

This eigenvalue structure indicates that for any positive position $x(t) > 0$ of the sliding mass, the system exhibits hyperbolic behavior around the equilibrium point $(\phi, \iota) = (0, 0)$. Specifically, the equilibrium is a saddle point with one stable and one unstable direction. This means that the pendulum in its downward position does not possess Lyapunov stability; almost all initial perturbations will cause the system to diverge from this equilibrium.

The physical interpretation is that the pendulum with a sliding mass, even when initially positioned near its downward equilibrium, will tend to swing away from this position over time rather than return to it. This instability is a direct consequence of the coupling between the sliding mass position and the angular dynamics of the pendulum.

3.8 *Oscillations in the Vertical Position*

Having established the instability of the equilibrium at $\phi = 0$, we now investigate whether oscillatory motion around this equilibrium is possible. For $x \geq 0$ and $\phi = 0$, the pendulum is in the vertical downward position. When $\iota = 0$ as well, the system is in a state of equilibrium. The question is whether small perturbations can lead to oscillations around this position.

We again examine the angular dynamics governed by

$$\begin{cases} \dot{\phi} = \frac{\iota}{mx^2+J} \\ \dot{\iota} = mgx\sin(\phi). \end{cases}$$

Let us consider the case when ϕ and ι have the same sign. Suppose that at some time t':

$$\begin{cases} \phi(t') & > 0 \\ \iota(t') & > 0. \end{cases}$$

Given that $x(t) \geq 0$ for all $t \in \mathbb{R}$ and the physical parameters m, g, and J are all positive, we can determine the signs of the derivatives at t':

$$\begin{cases} \dot{\phi}(t') & = \frac{\iota(t')}{mx^2(t')+J} > 0 \\ \dot{\iota}(t') & = mgx(t')\sin(\phi(t')) \geq 0. \end{cases}$$

These inequalities imply that for $t \geq t'$:

$$\begin{cases} \dot{\phi}(t) & > 0 \\ \iota(t) & \geq 0. \end{cases}$$

This means that $\phi(t)$ will monotonically increase for all $t \geq t'$. There is no possibility for ϕ to reverse direction and return toward zero, which would be necessary for oscillatory behavior. Therefore, the pendulum cannot exhibit oscillations of the angle ϕ in the vicinity of 0 for any trajectory where $x(t) \geq 0$.

An analogous argument applies when both $\phi(t')$ and $\iota(t')$ are negative. In this case, $\phi(t)$ will monotonically decrease for $t \geq t'$, again preventing oscillatory behavior.

This reasoning also holds for the case where

$$\begin{cases} \phi(t') & = 0 \\ \iota(t') & \neq 0. \end{cases}$$

In this scenario, the angle will immediately begin moving away from zero and continue in that direction.

More generally, whenever either:

$$\phi(t') \geq 0 \quad \text{and} \quad \iota(t') > 0$$

or

$$\phi(t') \leq 0 \quad \text{and} \quad \iota(t') < 0$$

the angle ϕ will monotonically move away from zero, prohibiting oscillatory behavior.

This mathematical analysis confirms what we already suspected from the stability analysis: the equilibrium at $\phi = 0$ is not only unstable but specifically acts as a saddle point from which trajectories diverge. The physical interpretation is that a pendulum with a sliding mass positioned at any positive distance from the pivot cannot sustain oscillations around the downward vertical position. Instead, once perturbed, it will rotate continuously in one direction.

3.9 *Asymptotic Stability*

Let us now consider a special case where the sliding mass is located at a significant distance from the pivot, specifically $x \geq a > 0$ where a is chosen such that $ma^2 \gg J$. Under this condition, the sliding mass can be approximated as a point mass, and the moment of inertia term becomes negligible: $J \approx 0$.

With this simplification, when we apply the algorithm for constructing the vector field, the angular dynamics reduce to

$$\begin{cases} \dot{\phi} = \frac{\iota}{mx^2} \\ \dot{\iota} = mgx\sin(\phi). \end{cases}$$

The goal is to determine whether we can impose a constraint that would create asymptotic stability for the pendulum in the vertical position. Let (ϕ_0, ι_0) represent the initial angular position and momentum, with $\phi_0 > 0$ and $\iota_0 > 0$, and let x_0 be the initial position of the sliding mass on the rod.

For small angular displacements, we can use the approximation $\sin(\phi) \approx \phi$, which yields

$$\begin{cases} \dot{\phi} = \frac{\iota}{mx^2} \\ \dot{\iota} = mgx\phi. \end{cases}$$

From this linearized system, we can derive a second-order differential equation for ϕ:

$$\ddot{\phi} + 2\frac{\dot{x}}{x}\dot{\phi} - \frac{g}{x}\phi = 0.$$

Let us consider a solution of the form:

$$\phi(t) = k_1 \exp(-k_2 t),$$

where $k_1 = \phi_0$ and $k_2 = -\frac{\iota_0}{mx_0^2\phi_0}$.

Substituting this proposed solution into the differential equation for ϕ, we obtain an equation for the position function $x(t)$:

$$k_2^2 x - 2k_2\dot{x} = g.$$

Solving this differential equation yields

$$x(t) = c\exp\left(\frac{k_2}{2}t\right) + \frac{g}{k_2^2}.$$

The constant c can be determined from the initial condition $x(0) = x_0$:

$$c = x_0 - \frac{g}{k_2^2}.$$

From the uniqueness of the solution to the Cauchy problem, we can conclude that given such a function $x(t)$ with initial conditions (ϕ_0, ι_0) and $x_0 > \frac{g}{k_2^2}$, we will obtain an asymptotically stable solution approaching the equilibrium point $(0, 0)$.

To implement this solution as a constraint, we need to determine the corresponding momentum function. Using the Legendre transformation, we have

$$p = m\dot{x} = md\pi \circ dx(\dot{x}, \dot{\phi}, \dot{p}, \dot{\iota}) = \frac{mk_1}{2}\left(x - \frac{g}{k_2^2}\right).$$

Therefore, by imposing the nonholonomic constraint:

$$f(x) = \frac{mk_1}{2}\left(x - \frac{g}{k_2^2}\right)$$

with the specified initial conditions, we can achieve asymptotic stability with respect to the angular position ϕ.

This result demonstrates that although the vertical position is naturally unstable for a pendulum with a sliding mass, we can design a specific nonholonomic constraint on the sliding mass movement that stabilizes this equilibrium. This approach has potential applications in control systems where stabilization of inherently unstable equilibria is required.

4 Inverted Pendulum

The inverted pendulum is one of the most widely studied benchmark problems in control theory, with applications spanning numerous engineering disciplines. This seemingly simple system, characterized by an inherently unstable equilibrium that requires active control strategies or specialized mechanical constraints to achieve stability, serves as an ideal case study for exploring various stabilization mechanisms, including nonholonomic constraints. Segway-type mobile systems based on the inverted pendulum principle have demonstrated commercial viability not only in personal transportation but also in logistics applications [8]. Furthermore, the fundamental principles of inverted pendulum control can be translated to diverse real-world engineering challenges, including robotic balancing systems, walking assistance devices for patients, object transport drones, and even modeling approaches for large-scale structures like buildings [9]. The adaptability of these control principles extends even to aerospace engineering. NASA's Space Launch System's Adaptive Augmenting Control algorithm, originally designed for aerodynamically unstable rockets, has been validated on pole-cart platforms due to the fundamental dynamic similarities between these seemingly different systems [10].

In this section, we focus on a special case, the vertically-driven inverted pendulum, which demonstrates a remarkable physical phenomenon where rapid oscillations can stabilize an inherently unstable equilibrium position. This phenomenon was first discovered and demonstrated in the early twentieth century [11], rediscovered in the mid-twentieth century [12], and thoroughly investigated in the late twentieth century [15, 16]. The system consists of a rigid pendulum whose suspension point undergoes high-frequency vertical oscillations. When these oscillations have sufficient frequency and amplitude, the normally unstable inverted position becomes dynamically stabilized. This counterintuitive behavior occurs because the average effect of the oscillations creates an effective potential minimum at the inverted position. This dynamic stabilization phenomenon has applications in various fields, from Paul traps for charged particles [13] to the stability analysis of mechanical systems [14]. The vertically-driven inverted pendulum provides an elegant demonstration of how time-dependent constraints can fundamentally alter a system's stability properties.

Throughout the analysis, we will derive the Hamiltonian formulation of the inverted pendulum, introduce specific nonholonomic constraints representing the vertical oscillations, construct the appropriate vector fields, and perform stability analysis. By deriving the classical Mathieu equation from the geometric approach, we establish a bridge between the theoretical framework and well-established results in the physics literature. This analysis not only validates the methodology but also provides insights into designing constraint-based control strategies for unstable mechanical systems. The connection between the geometric framework developed in previous sections and practical control applications demonstrates the power of this approach for understanding and controlling complex dynamic systems.

4.1 Vertically-Driven Inverted Pendulum

Consider a planar mathematical pendulum whose suspension point undergoes vertical oscillations. This classical mechanical system exhibits rich dynamical behavior and serves as an important model for understanding stabilization through time-dependent constraints. Let us establish the fundamental parameters of the system: the pendulum consists of a rigid, massless rod of length l with a bob of mass m attached to its end. The position of the suspension point is measured along the vertical axis (oriented upwards) and denoted by the variable y. The angular displacement of the rod from the vertical axis is represented by ϕ, measured in the counterclockwise direction, see Fig. 2.

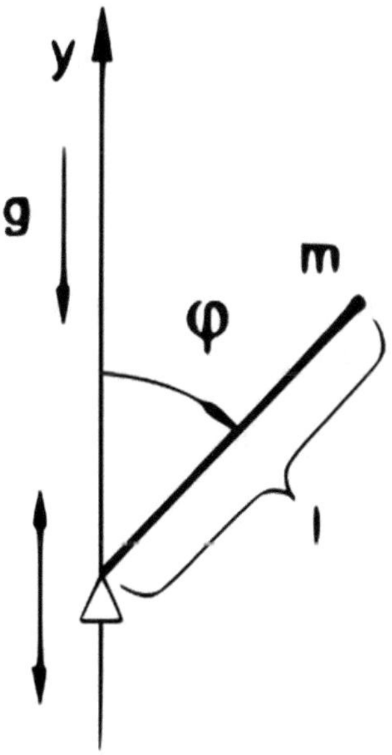

Fig. 2 Vertically-driven inverted pendulum

With these parameters defined, the configuration space of the system can be formally represented as

$$M = \{(\phi, y) \mid \phi \in S^1, y \in \mathbb{R}\}.$$

This configuration space forms a cylinder, as ϕ is periodic with period 2π while y can take any real value. For analysis, we will work within the chart

$$U = \{b \mid \phi(b) \in (-\frac{\pi}{2}, \frac{\pi}{2})\},$$

which encompasses the region where the pendulum is less than 90°C from the vertical in either direction.

The kinetic energy of the system arises from both the rotational motion of the pendulum and the vertical motion of the suspension point. Due to the coupling between these motions, the kinetic energy includes a cross-term:

$$G|_{(\phi,y)}(\dot{\phi}, \dot{y}) = \frac{ml^2}{2}\dot{\phi}^2 - ml\sin(\phi)\dot{\phi}\dot{y} + \frac{m}{2}\dot{y}^2.$$

The first term represents the rotational kinetic energy of the pendulum bob, the third term accounts for the vertical kinetic energy of the suspension point, and the middle term captures the coupling between these two motions.

The potential energy of the system is determined by the height of the pendulum bob and the suspension point against gravity:

$$V(\phi, y) = mg(l\cos(\phi) + y).$$

Here, $l\cos(\phi)$ gives the vertical component of the pendulum bob's position relative to the suspension point, while y represents the height of the suspension point itself.

4.2 Hamiltonian Formulation

To develop a Hamiltonian description of the system, we must transition from the Lagrangian variables $(\phi, y, \dot{\phi}, \dot{y})$ to the canonical phase space coordinates (ϕ, y, ι, q). This transformation is achieved through the Legendre transformation, which defines the generalized momenta:

$$\iota = \frac{\partial G}{\partial \dot{\phi}} = ml^2\dot{\phi} - ml\sin(\phi)\dot{y}$$
$$q = \frac{\partial G}{\partial \dot{y}} = m\dot{y} - ml\sin(\phi)\dot{\phi}.$$

Here, ι represents the angular momentum conjugate to ϕ, while q is the linear momentum conjugate to y. Notice that due to the coupling in the kinetic energy, each momentum depends on both velocities, reflecting the interconnected nature of the pendulum's dynamics.

In these canonical coordinates, the symplectic structure of the phase space $\Phi = T^*M$ is represented by the canonical 2-form $\Omega = d\iota \wedge d\phi + dq \wedge dy$. The kinetic energy, when expressed in terms of these phase space variables, takes the form:

$$\tilde{G}(\phi, y, \iota, q) = \frac{1}{2ml^2\cos^2(\phi)}(\iota^2 + 2l\sin(\phi)\iota q + l^2q^2).$$

This expression reveals how the pendulum's dynamics become more complex near $\phi = \pm\frac{\pi}{2}$, where $\cos(\phi)$ approaches zero, potentially leading to singular behavior.

The complete dynamics of the system is governed by the generalized Hamiltonian one-form:

$$E = dH = d(\tilde{G} + V \circ \pi),$$

where $\pi : \Phi \to M$ is the natural projection from the phase space to the configuration space. This Hamiltonian formulation provides the foundation for analyzing the system's behavior under various constraints, particularly the oscillatory motion of the suspension point that we will introduce in subsequent sections.

4.3 Non-smooth Nonholonomic Constraints for the Oscillating Suspension

To model the vertical oscillations of the pendulum's suspension point, we introduce a piecewise-defined nonholonomic constraint that governs the vertical velocity. This constraint is non-smooth due to its directional dependence, creating different dynamics during the upward and downward phases of the oscillation:

$$\begin{cases} \dot{y} - f(y) = 0 & \text{if } \dot{y} \geq 0 \\ \dot{y} - g(y) = 0 & \text{if } \dot{y} \leq 0, \end{cases}$$

where $f, g \in C^\infty((y_0, y_1))$ are smooth functions defined on the interval (y_0, y_1), and they satisfy boundary conditions $f(y_0) = f(y_1) = g(y_0) = g(y_1) = 0$. These conditions ensure that the suspension point oscillates between the positions y_0 and y_1, with the velocity vanishing at the endpoints of the oscillation.

The non-smoothness occurs at transition points where $\dot{y} = 0$, as the system switches between two different constraint manifolds. This switching between constraints represents the physical change in direction of the suspension point's motion and requires the geometric framework for non-smooth constraints that we developed earlier.

To incorporate these constraints into the Hamiltonian framework, we must express the vertical velocity $\dot{y}$ in terms of the canonical phase space variables. Using the inverse Legendre transformation, we obtain

$$\dot{y} = \frac{lq + \sin(\phi)\iota}{ml\cos^2(\phi)}.$$

This expression reveals the coupling between the linear momentum q, the angular momentum ι, and the angular position ϕ. Note that this expression becomes singular when ϕ approaches $\pm\frac{\pi}{2}$, which is why we restricted our analysis to the chart

$$U = \{b \mid \phi(b) \in (-\frac{\pi}{2}, \frac{\pi}{2})\}.$$

For the case when $\dot{y} > 0$ (upward motion of the suspension point), we construct a submanifold in the phase space corresponding to the nonholonomic constraint:

$$\Pi^+ = \{(\phi, y, \iota, q) \in \Phi \mid F(\phi, y, \iota, q) = 0, y \in (y_0, y_1)\},$$

where the constraint function F is defined as

$$F(\phi, y, \iota, q) = lq + \sin(\phi)\iota - f(y)ml\cos^2(\phi).$$

This constraint function equates the vertical velocity $\dot{y}$ with the prescribed function $f(y)$ when the suspension point is moving upward. A similar construction would define the submanifold Π^- for the downward motion using the function $g(y)$.

4.4 Constrained Vector Field

To determine the dynamics of the constrained system, we must construct a vector field on the submanifold Π^+ that respects both the Hamiltonian structure and the nonholonomic constraint. Following the geometric approach, we proceed through several steps.

First, we construct the tangent space $T_b\Pi^+$ at a point $b \in \Pi^+$:

$$T_b\Pi^+ = \{\xi \in T_b\Phi \mid dF|_b(\xi) = 0\}.$$

The differential of the constraint function is given by

$$\begin{aligned} dF &= l\,dq + \sin(\phi)\,d\iota - ml\cos^2(\phi) f'(y)\,dy \\ &\quad + (\cos(\phi)\iota + 2ml\cos(\phi)\sin(\phi) f(y))\,d\phi. \end{aligned}$$

Next, we construct an affine subspace $T_b^0\Pi^+ \subset T_b\Pi^+$ consisting of vectors that satisfy the nonholonomic constraint at the velocity level:

$$\begin{aligned} T_b^0\Pi^+ &= \{\xi \in T_b\Pi^+ \mid \sigma|_b(\xi) - ml\cos^2(\phi) f(y) = 0\} \\ &= \{\xi \in T_b\Phi \mid dF|_b(\xi) = 0, \sigma|_b(\xi) - ml\cos^2(\phi) f(y) = 0\}, \end{aligned}$$

where $\sigma = ml\cos^2(\phi)\,dy$ is a covector field on Φ.

To facilitate analysis, we construct a corresponding linear subspace $T_b^1\Pi^+$ by shifting the affine subspace $T_b^0\Pi^+$:

$$T_b^1\Pi^+ = \{\xi \in T_b\Phi \mid dF|_b(\xi) = 0, \sigma|_b(\xi) = 0\}.$$

Following the general framework, we construct a covector $\omega|_b \in T_b^{1^*}\Pi^+$ that corresponds to the image of the generalized Hamiltonian $E \in \Lambda(\Phi)$ under the factorization $T_b^*\Phi/\{dF|_b = 0, \sigma|_b = 0\}$.

From this covector, we construct the vector $X^0|_b \in T_b^0\Pi_+$ using:

$$X^0|_b = I|_b(i_{I|_b(E|_b + \mu(b) dF|_b + \lambda(b)\sigma|_b)}\Omega|_b).$$

For this vector to lie in $T_b^0\Pi_+$, the parameters $\mu(b)$ and $\lambda(b)$ must satisfy

$$\begin{aligned} dF|_b(X^0|_b) &= 0 \\ \sigma|_b(X^0|_b) - ml\cos^2(\phi) f(y) &= 0. \end{aligned}$$

Solving these conditions, we obtain

$$\begin{aligned} \mu(b) &= 0 \\ \lambda(b) &= \frac{1}{ml^2\cos^2(\phi)}\left(\frac{\iota + l\sin(\phi)q}{\cos(\phi)} - f'(y)(lq + \sin(\phi)\iota)\right) - \frac{g}{l}. \end{aligned}$$

With these values, we can now explicitly describe the vector field over Π^+. After performing the necessary calculations, we arrive at the following system of differential equations that govern the dynamics of the constrained pendulum:

$$\begin{cases} \dot{\phi} &= \dfrac{\iota + l\sin(\phi)q}{ml^2\cos^2(\phi)} \\ \dot{y} &= \dfrac{lq + \sin(\phi)\iota}{ml\cos^2(\phi)} \\ \dot{\iota} &= mgl\sin(\phi) - \dfrac{(\iota + l\sin(\phi)q)(lq + \sin(\phi)\iota)}{ml^2\cos^3(\phi)} \\ \dot{q} &= f'(y)\left(q + \dfrac{\sin(\phi)}{l}\iota\right) + \dfrac{(\iota + l\sin(\phi)q)^2}{ml^3\cos^3(\phi)} - mg\sin^2(\phi). \end{cases}$$

This system fully describes the constrained dynamics of the vertically-driven inverted pendulum during the upward motion of the suspension point. An analogous construction can be carried out for the downward motion by employing the constraint function $g(y)$.

At the transition points where $\dot{y} = 0$, special consideration is needed to ensure the continuity of the motion, applying the framework for non-smooth constraint transitions developed earlier. The complete dynamics of the system combines the vector fields from both regimes into a piecewise-smooth vector field on the full constraint manifold $\Pi = \Pi^+ \cup \Pi^-$.

4.5 Stability Analysis

From the system of equations derived in the previous subsection, we can now extract a second-order differential equation that describes the angular motion of the pendulum. Taking appropriate derivatives of the system equations and eliminating variables, we arrive at the following equation for the angular coordinate ϕ:

$$\ddot{\phi} = \left(g + f'(y)\dot{y}\right)\frac{\sin(\phi)}{l}.$$

This equation reveals how the effective acceleration of the pendulum depends not only on gravity but also on the motion of the suspension point through the term $f'(y)\dot{y}$. The oscillation of the suspension point effectively modulates the gravitational acceleration, which is the key mechanism behind the potential stabilization of the inverted position.

To model the specific case of harmonic vertical oscillations, the classical system extensively studied in the physics literature, we define the constraint functions for the motion of the suspension point as

$$\begin{cases} \dot{y} - \gamma\sqrt{a^2 - y^2} = 0 & \text{if } \dot{y} \geq 0 \\ \dot{y} + \gamma\sqrt{a^2 - y^2} = 0 & \text{if } \dot{y} \leq 0 \end{cases}$$

with the initial condition $y(0) = -a$. This particular choice of constraint functions creates simple harmonic motion of the suspension point, with amplitude a and frequency γ. The solution to these constraint equations is

$$y(t) = a\sin(\gamma t)$$

and consequently:

$$\dot{y}(t) = a\gamma\cos(\gamma t).$$

Substituting these expressions into the differential equation for ϕ and using the small-angle approximation $\sin(\phi) \approx \phi$ (valid near the inverted position), we obtain

$$\ddot{\phi} = \left(g + a\gamma^2\cos(\gamma t)\right)\frac{\phi}{l}.$$

This is precisely the Mathieu equation, a canonical form of a differential equation with time-periodic coefficients. The Mathieu equation plays a fundamental role in the theory of parametric resonance and appears in various physical contexts beyond the inverted pendulum, including the stability of floating bodies, quantum mechanical systems, and certain electromagnetic wave phenomena [17].

The Mathieu equation derived above has solutions characterized by special functions known as Mathieu functions. These solutions exhibit a rich structure of stable and unstable regions in parameter space, determined by the values of the amplitude a, frequency γ, pendulum length l, and gravitational acceleration g.

For the inverted pendulum system, the critical question is whether the upright position ($\phi = 0$) can be stabilized by the vertical oscillations of the suspension point. The classical theory of the Mathieu equation provides us with a stability criterion: the inverted position becomes dynamically stable when:

$$a\gamma^2 > 2g.$$

This remarkable condition reveals that stability is achieved when the maximum acceleration of the suspension point ($a\gamma^2$) exceeds twice the gravitational acceleration. This counterintuitive result demonstrates how rapid oscillations can effectively invert the potential energy landscape of the system, creating a local minimum at the otherwise unstable equilibrium point.

The physical mechanism behind this stabilization can be understood through time-scale separation. When the oscillation frequency is sufficiently high, the pendulum experiences the average effect of the oscillation rather than responding to each individual cycle. The rapid vertical motion creates an effective potential that, under the right conditions, forms a "potential well" at the inverted position, trapping the pendulum near the unstable equilibrium.

This phenomenon of dynamic stabilization illustrates a profound principle in nonlinear dynamics: a system can be stabilized not only through static forces but also through appropriate time-dependent perturbations. The inverted pendulum with an oscillating suspension point serves as one of the most elegant demonstrations of this principle, showing how instability can be overcome through carefully designed dynamic constraints.

5 Conclusion

In this work, we have demonstrated a comprehensive geometric framework for analyzing mechanical systems with nonholonomic constraints using Hamiltonian mechanics [4, 6]. This approach provides a systematic method for constructing vector fields that govern the dynamics of constrained systems, offering both mathematical elegance and practical utility for understanding complex mechanical behaviors.

For a given nonholonomic constraint imposed on a mechanical system, we have shown how to construct a submanifold of the phase space that represents all admissible states. The canonical two-form restricted to this submanifold, together with the carefully constructed distribution in the tangent bundle, defines a symplectic structure that allows us to determine a vector field corresponding to the system's Hamiltonian. The trajectories along this vector field provide a complete description of the system's motion, yielding valuable insights into its behavior.

The power of this approach has been demonstrated through detailed analyses of two mechanical systems: the pendulum with a sliding mass and the inverted pendulum with an oscillating suspension point. These examples illustrate how the geometric framework can uncover profound dynamical properties, including stability characteristics and control possibilities. For the sliding mass pendulum, we established conditions under which the vertical position exhibits instability, and furthermore, identified specific nonholonomic constraints that can provide asymptotic stability. In the case of the inverted pendulum, we derived the Mathieu equation and the corresponding stability criterion $a\gamma^2 > 2g$ directly from the geometric formulation, demonstrating that rapid oscillations of the suspension point can stabilize an otherwise unstable equilibrium [15].

A significant contribution of this work is the extension of the Hamiltonian framework to handle non-smooth nonholonomic constraints, which arise naturally in mechanical systems with directional dependencies, impacts, and sliding phenomena. By carefully analyzing the behavior at constraint transitions, we have developed a methodology that provides a unified treatment of both smooth and discontinuous dynamics. This extension is particularly valuable for modeling realistic mechanical systems where idealized smooth constraints are insufficient.

The framework presented here has broad implications for control theory and mechanical engineering. The ability to stabilize inherently unstable equilibria through properly designed nonholonomic constraints offers a powerful tool for control system design [22]. Moreover, the approach to non-smooth constraints provides a rigorous foundation for analyzing systems with impacts, friction, and other discontinuous phenomena that are ubiquitous in practical applications.

Future research could extend this framework in several promising directions. First, the methodology could be applied to more complex mechanical systems with multiple degrees of freedom and coupled constraints. Second, the connection between the geometric approach and optimal control theory deserves further investigation, particularly for systems with non-smooth dynamics where traditional control techniques may be insufficient. Finally, the extension to quantum mechanical systems with constraints represents an intriguing possibility for bridging classical and quantum mechanics within a unified geometric framework.

In conclusion, the geometric Hamiltonian approach to nonholonomic mechanical systems presented in this work provides a powerful theoretical framework with clear practical applications. By analyzing the symplectic structure underlying constrained dynamics and demonstrating its application to classical pendulum systems, we have shown how geometric insights can lead to a deeper understanding of mechanical behavior and novel approaches to control and stabilization.

References

1. F. Klein, *Lectures on Mathematics*. An Imprint of the American Mathematical Society (AMS Chelsea Publishing, 2000)
2. E. Fermi, Notes on Quantum Mechanics (University of Chicago Press, 1995)
3. A. Vinogradov, I. Krasil'shchik, What is the Hamiltonian formalism. London Math. Soc. Lecture Notes Ser. **60**, 241–266 (1981)
4. A. Vinogradov, B. Kupershmidt, The structure of Hamiltonian mechanics. London Math. Soc. Lect. Notes Ser. **60**, 173–239 (1981)
5. J. Nestruev, Smooth Manifolds and Observables (Springer, Berlin, 2003)
6. J. Cortés, A. Vinogradov, Hamiltonian theory of constrained impulsive motion. J. Math. Phys. **47** (2006)
7. F. Pugliese, A. Vinogradov, Geometry of inelastic collisions. Acta Applicandae Mathematica **72**, 77–85 (2002)
8. C. Zauner, A. Müller, H. Gattringer, M. Jörgl, Trajectory planning for a Segway model exploiting inherent feedforward structure. IFAC-PapersOnLine **53**, 8444–8449 (2020)
9. K. Ioannis, M. Lazaros, Inverted pendulum: a system with innumerable applications, in *9th International Week Dedicated to Maths* (2017)
10. J. Pei, P. Rothhaar, Demonstration of the space launch system augmenting adaptive control algorithm on pole-cart platform, in *2018 AIAA Guidance, Navigation, And Control Conference* (2018), p. 0608
11. A. Stephenson, On an induced stability. Phil. Mag. **15**, 233–236; On a new type of dynamical stability Mem. Proc. Manch. Lit. Phil. Soc. **52**, 1–10 (1908)
12. P. Kapitza, Collected Papers of P. L Kapitza, in Haar, D.T. (ed.) (Pergamon Press, Oxford, 1964)
13. W. Paul, Electromagnetic traps for charged and neutral particles. Rev. Mod. Phys. **62**, 531 (1990)
14. M. Levi, Stability of the inverted pendulum–a topological explanation. SIAM Rev. **30**, 639–644 (1988)
15. L. Landau, E. Lifshitz, Mechanics Pergamon, p. 156 (New York, 1976)
16. V. Arnold, Ordinary Differential Equations (Springer Science & Business Media, 1992)
17. L. Ruby, Applications of the Mathieu equation. Am. J. Phys. **64**, 39–44 (1996)
18. V. Aslanov, Stability of a pendulum with a moving mass: the averaging method. J. Sound Vib. **445**, 261–269 (2019)

19. D. Stilling, W. Szyszkowski, Controlling angular oscillations through mass reconfiguration: a variable length pendulum case. Int. J. Non-Linear Mech. **37**, 89–99 (2002)
20. G. Yakubu, P. Olejnik, J. Awrejcewicz, On the modeling and simulation of variable-length pendulum systems: a review. Arch. Comput. Methods Eng. **29**, 2397–2415 (2022)
21. O. Gutiérrez-Frías, J. Martí1nez-García, R. Garrido-Moctezuma, Stabilization on a physical pendulum with moving mass, in *Advances in Computational Intelligence* (2009), pp. 509–518
22. A. Bloch, A. Bloch, Nonholonomic Mechanics (Springer, Berlin, 2015)
23. W. Szyszkowski, D. Stilling, On damping properties of a frictionless physical pendulum with a moving mass. Int. J. Non-Linear Mech. **40**, 669–681 (2005)

MIX
Papier aus verantwortungsvollen Quellen
Paper from responsible sources
FSC® C105338

If you have any concerns about our products,
you can contact us on
ProductSafety@springernature.com

In case Publisher is established outside the EU,
the EU authorized representative is:
Springer Nature Customer Service Center GmbH
Europaplatz 3, 69115 Heidelberg, Germany

Printed by Libri Plureos GmbH
in Hamburg, Germany